Yao Ngoran Bazin

Dreaming of local development

Yao Ngoran Bazin

Dreaming of local development

Towards a bolder, more sustainable commitment

ScienciaScripts

Imprint
Any brand names and product names mentioned in this book are subject to trademark, brand or patent protection and are trademarks or registered trademarks of their respective holders. The use of brand names, product names, common names, trade names, product descriptions etc. even without a particular marking in this work is in no way to be construed to mean that such names may be regarded as unrestricted in respect of trademark and brand protection legislation and could thus be used by anyone.

Cover image: www.ingimage.com

This book is a translation from the original published under ISBN 978-3-639-54580-7.

Publisher:
Sciencia Scripts
is a trademark of
Dodo Books Indian Ocean Ltd. and OmniScriptum S.R.L publishing group

120 High Road, East Finchley, London, N2 9ED, United Kingdom
Str. Armeneasca 28/1, office 1, Chisinau MD-2012, Republic of Moldova, Europe
Managing Directors: Ieva Konstantinova, Victoria Ursu
info@omniscriptum.com

Printed at: see last page
ISBN: 978-620-3-50303-6

Table of contents

Introduction

Sixty (60) years after Côte d'Ivoire gained independence, the country is still grappling with problems linked to the provision of urban services and basic infrastructure, and, more generally, with the impossibility of improving the living conditions and environment of the population, in line with the sixteen (16) responsibilities or competencies transferred to local authorities. The truth is that Côte d'Ivoire has not yet succeeded in laying the real foundations for local development. Côte d'Ivoire's current decentralization system is a mess. The country has undeniable potential for progress. But it is not succeeding. There is no doubt that it has failed.

The provision of basic services such as water, electricity, schools, health and education is an essential challenge for local authorities in Côte d'Ivoire. There are other challenges facing the nation. The transformation of raw materials into finished or semi-finished products, the development of economic infrastructures, the improvement of the quality of the physical environment of human settlements, youth unemployment, the insecurity of people and their property, the development of education and research, are also concerns of the national community. These questions continue to be acute, because those in power most often lack the courage to question the options or approaches adopted, even when they have not enabled Côte d'Ivoire to achieve the specific objectives sought at the level of each skill or responsibility.

Our national decentralization policy is not working. As a result, the much-vaunted local development cannot take place, due to the existence of numerous barriers. Côte d'Ivoire must agree to review its entire decentralization policy, if it wants its march towards progress to be a tangible and lasting reality.

In the 16 areas covered by decentralization in Côte d'Ivoire, there are no clearly defined sectoral policies, with precise roles and responsibilities for the players involved. The conduct of decentralization in Côte d'Ivoire is still marked by the omnipresence of the State and its branches. This has forced many players to remain inactive where they could have played an active and decisive role in grassroots development. Instead, they all remained on the sidelines. For example, when it comes to keeping our neighborhoods, villages and towns clean, Côte d'Ivoire is struggling, because the approaches adopted to date have been totally unsuitable. The national policy for waste management and sanitation in the country's human settlements does not involve the population. It's not just the population that's concerned, but also social work. Taking social work into account is an element to be taken into account in the formulation and implementation of a genuine solid waste management and sanitation policy in Côte d'Ivoire. The filth and detritus that greet us at the entrance to every neighborhood, every village and every town in the country are nothing more than the manifestation of the failure of solid waste management and the sanitation of the country's human settlements.

One of the main aims of our work is to examine how far Côte d'Ivoire has come or gone in its quest for local development. This examination is not that of officials or politicians, local decision-makers or elected representatives. It is that of a citizen concerned about the dismal state of local authorities as a whole. Our course of action is not simply to point out the weaknesses of the system. We need to go beyond our observations to propose solutions that can help the country's local authorities move forward. These days, we're talking about shaping the future of local authorities and their populations. Defining the future of local authorities, their towns and their inhabitants starts with an uncompromising assessment of the situation.

I wanted to write a book about local development in Côte d'Ivoire. The purpose of this book is to help bridge the country's local development gap. We talk a lot about local development, but we don't always know which document to refer to. In fact, there is no book to refer to when formulating policy, defining development programs or projects, and implementing and evaluating them.

Côte d'Ivoire's local councillors must admit that running a community is no easy task. There are many difficulties involved in exercising the responsibilities and competencies entrusted to them. In the face of these difficulties, whether known or unknown, they must organize or reorganize themselves in order to progressively achieve the hoped-for results, particularly in terms of improving people's living conditions and environment. They need to earn the trust of stakeholders, by demonstrating efficiency to give their team credibility, and by being more aggressive and achieving a good level of competitiveness.

The book is about the local development that I dream of, and that other citizens certainly dream of. This local development is not what we see today. Our thought is that, from what we all see today, we can make the continuous improvements that are needed, without complacency, to offer local development more in line with Côte d'Ivoire's ambition.

Part 1: UNDERSTANDING LOCAL DEVELOPMENT

Chapter 1: Introduction to local development

At the outset of this book, I feel it's necessary to propose an introductory chapter, designed to provide a clear understanding of the concept of local development. From this understanding will certainly flow our willingness and commitment to act for the smooth running of the country's local authorities.

1.1. From centralization to decentralization of power management

Until 1980, Côte d'Ivoire's approach to development was fairly centralized. This approach was characterized by a very strong concentration by the State of the powers to think, decide and act. This approach left no room for manoeuvre for local populations or communities. This approach had certainly produced results. But it still failed to live up to the stated ambitions of those in power and the deep-seated aspirations of the people. Central government therefore decided to allow local players to play a greater role in the development of the regions and, by extension, the country. In many respects, the move from central to local government is a well-considered choice.

Initially, the State concentrated the nation's elites and all development policy was formulated from Abidjan, the capital. By 1980, the regions already had a fairly large number of human resources, because Côte d'Ivoire's efforts in the field of executive training had produced convincing results. It should be remembered that Côte d'Ivoire lagged far behind in education and management training at independence in 1960. Twenty years after Côte d'Ivoire gained independence, executives had been trained in most areas of human activity. These executives had demonstrated their ability to take over from expatriates in defining development programs and projects. But there was no real political will to empower local elites.

In 1980, the authorities felt that Côte d'Ivoire could embark on a genuine decentralization policy. A major political milestone was reached with the decision to launch a vast movement to decentralize the management of power in the country.

1.2. Defining the concept of local development

What is our understanding of the concept of "local development"? The question is not without meaning. For us, local development or locally-based development is development that is designed, organized, implemented and evaluated in a given territorial area, with the full, effective and sustainable participation of local populations and the mobilization of appropriate resources.

Local development is a paradigm, a way of thinking and acting. It is not a static concept. It is dynamic. This dynamic manifests itself through space, demography, culture and people's projects.

Our decentralization, and the local development it should engender, is not complete. Against this backdrop of failure, we are obviously seeing more complaints and lamentations than satisfaction among the population. We must have the courage and humility to recognize that our local development needs to be built in Côte d'Ivoire.

1.3. Local development objectives

The general aim of local development, to put it more simply, is to make a given locality or region, with its strengths and weaknesses, a good place to live. A place where local populations can freely build themselves, individually or collectively, through the realization of their dreams and projects.

To achieve local development, or to move towards real and effective local development, Côte d'Ivoire needs to propose innovations in the general system of governance of terroirs and territories. Local development has the following specific objectives:

- Getting people to accept to question themselves;
- Understand local development to act more effectively and achieve results in line with expectations;
- Contribute to a better understanding of the roles and responsibilities of local elected officials and their various partners;
- Contribute to preparing the rules of the game for the competition in which local authorities should participate;
- Empowering local populations for a better implementation of development actions;
- Promote knowledge, know-how, know-how to do and know-how to be, among local populations.

1.4. Understanding local development

People's understanding of the concept of local development depends to a large extent on their level of education. However, education alone is not enough to make an enlightened citizen of local development issues. In Ivorian society, there are many citizens in all regions and therefore in all cultural areas of the country, including illiterate people, who have understood the need to carry out development projects or actions in their locality and in different sectors of human activity. They may be illiterate, but they have achieved feats of vision and understanding of development in their home localities that few citizens, even those with university qualifications, have been able to achieve.

In some parts of the country, these illiterates were ahead of the game. Wherever there was a lack of infrastructure or social and economic facilities, they proposed solutions to the community. For example, they proposed the opening up of a road to open up one or more localities, built a school to prepare the locality for the future, enrolled all their children in school, moved a village from an unsuitable site to another offering better development prospects. They have also created fish ponds for young people, and set up units to process certain agricultural products. They have provided leadership in their region. The list of these pioneers' achievements is not exhaustive.

These illiterate leaders didn't just work for their families. They were concerned with the interests of the community to which they belonged. They did local development without knowing it. Today, we can better appreciate the contribution of these illustrious illiterate pioneers to their community's march towards progress. Their commitment, their careers and their achievements have not gone unnoticed. They were role models at both local and national level. The merit of these men and women, educated or illiterate, is more than great.

The Ivorian population's understanding of the concept of local development is virtually identical in all cultural areas. In each cultural area, people have been carrying out worthwhile social, cultural, economic and even ecological activities for decades, without any outside intervention.

1.5. Beyond local development theories

There are many theories on development and local development. African governments, international cooperation organizations and economists or experts in development issues have tried to base their actions on theories or economic models tested in other political, spatial, socio-cultural or anthropological environments different from those of Africa. Their introduction and implementation, often without due care, in African countries have resulted in failure. Importing development models from outside the continent has not produced the desired results. In any case, they have not led to any real economic take-off, either locally or nationally. They have not laid the foundations for the endogenous, self-sustaining and sustainable development of African countries. Development specialists use these facts to explain the extraversion of Côte d'Ivoire's economy, for example.

All of which goes to show that local development paths for Côte d'Ivoire, for example, need to be explored, by questioning ourselves. We need to question ourselves. We have to start from who we are, what we have and what we want. These are the aspects that will enable us to create our own paths, namely those we can master. At any moment, we, the Ivorian people, can stop to see if things are working and if there are aspects we need to correct or change.

Today, one of the greatest dangers facing our society is that universities continue to teach things that are foreign to young people. We deplore the fact that these young people are poorly trained and ill-prepared, if at all, to face the challenges ahead. They are being trained to feed unemployment or to become civil servants, if they are not picked up along the way by politicians. At the end of their studies, most graduates of Côte d'Ivoire's public or private universities and grandes écoles know and can do nothing, even the simplest things, such as replacing a faulty light bulb in their parents' home. They are therefore incapable of participating in the fight against our country's underdevelopment. But we don't always ask ourselves what our universities are teaching. The political and academic authorities need to listen to us instead of calling us birds of ill omen. They need to review the content of the teaching offered to Ivorian youth. It is absolutely and imperatively necessary to undertake major reforms of our higher education system. For my part, I'll come back to the contribution of a university to local development in chapter 10.

To date, local development in Côte d'Ivoire has not yet produced appropriate results, due to decentralization falling short of Côte d'Ivoire's ambitions. An analysis of the situation reveals that the understanding of local development actions has not been sufficiently optimal. It is not effective at all levels of Ivorian society.

Chapter 2: Knowledge of the local authority

Our focus is on the local or territorial community, as the geographical space in which we intervene and carry out actions aimed at improving living conditions and environments. The implementation of development actions must be based on a good knowledge of the community. It therefore seems necessary to talk about the knowledge of the community on the part of the actors likely to carry out programs, projects or actions aimed at improving the well-being of the population.

2.1. Knowledge of the local community

2.1.1. Local authorities: a more or less harmonious whole

The local community is a whole, comprising natural resources (water, forest, hydrographic network, land resources) and plant species. Then there are the human resources, i.e. the people who live there, with their pasts and histories, their values, customs and traditions, their knowledge, know-how and interpersonal skills, their ambitions, projects and dreams. People must be in harmony, in balance with nature.

The local authority is a whole, with varied and more or less strong links between its various components. Knowledge of the local authority reveals the nature, intensity and even complexity of the relationships between its various components. Each local authority is different from the next, in every aspect. No two local authorities in Côte d'Ivoire are alike. The realities of one local authority are not those of another.

The local community is, or must be, in harmony with itself. This harmony is not immutable. It is dynamic, changeable, in line with the evolution of society and time. In fact, it can be disrupted by natural phenomena (droughts, storms, floods) or by human action, in particular mineral extraction, agricultural or pastoral activities, or national land-use planning and development. Human activities also include housing and infrastructure, in particular roads, freeways, hydroelectric dams, railroads, bridges, ports, airports, telecommunications facilities, etc.

2.1.2. Intrinsic knowledge of the local community

Any human intervention or action aimed at community development must be based on intrinsic knowledge of its realities. This knowledge requires patience, commitment and determination. The system or architecture of knowledge is also plural (geography, environment, history, culture, politics, sociology, anthropology, economics, technology, etc.). No area of human activity can be neglected. Knowledge of the history and anthropology of a locality helps to structure its identity, i.e. its specificity.

For the elected representatives of Ivorian local authorities, one of the actions to be undertaken in the process of their knowledge is to detect or identify opportunities to be transformed into programs, projects or actions for economic production and wealth creation. One of the key questions to be resolved by local councillors is precisely how to detect opportunities. Opportunities generally take two forms: the physical form they take, and the transformation they undergo, which translates into cause-and-effect actions. These actions need to be analyzed, finely and objectively, in order to facilitate decision-making.

To identify these opportunities and prepare development programs or projects, local authorities can call on experts from universities, research institutes or centers, or consultants with varied and specialized expertise in local development. Experts from these various regional or national institutions form a pool to which local elected representatives can turn when commissioning studies or work of a scientific, technical or organizational nature.

In the development process of a region or a nation, dynamic thinking is the raw material. This is based on research and technical studies, the products of which are used to facilitate decision-making by elected representatives and their partners. In this quest for knowledge of the community, national experts can be called upon to contribute. Before or after political independence, our leaders' systematic recourse to foreign expertise was justified by the virtual non-existence of their own skills.

Today, Côte d'Ivoire has made significant efforts to develop national expertise. Of course, when it comes to developing human resources, particularly in terms of quality, there's still a long way to go. Let's face it. To make Côte d'Ivoire's dream of development a reality, we need to create new, genuine universities that work, and offer high-quality research and teaching to students. This is one of the keys to development. To understand the need for this transition, we need to remember, for example, that the Republic of South Korea will have three hundred (300) public universities by 2020. Of course, comparison is not reason. But this gives us a sufficiently clear idea of the efforts Côte d'Ivoire needs to make in terms of human resources development and the promotion of science and technology.

National experts, from all disciplinary backgrounds, can be mobilized to participate in the definition of policies or programs, and in the design and implementation of development projects or actions at regional level. They have the advantage or merit of knowing relatively more about the physical geography, the people who populate the regions, their thoughts, their cultures, their behaviors and their activities. Through their detailed, objective analysis, they can help to understand the nature and evolution of the physical, chemical, socio-cultural, economic and environmental phenomena facing local communities. National experts can also support local elected representatives in their quest for development. It's worth pointing out that the presence of national experts alongside local elected representatives reassures development partners. It motivates them to work with them.

2.1.3. Statistics production and management

Good knowledge of local authorities must be based on, or supported by, the existence of viable statistics. Statistics are extremely important for any local authority, particularly for guiding and evaluating local action. They are also useful for forecasting local development.

When it comes to local development, one of the most common findings is the lack of statistics on what is being achieved in each community. Indeed, whether in terms of social services (water, electricity, schools, health), economic activity (formal and informal sectors combined), demographics or the environment, statistics are sorely lacking in Côte d'Ivoire's local authorities.

A detailed knowledge of statistics enables us to understand the manifestations of a phenomenon before taking action. They guide and facilitate the decisions that can support the actions to be taken. In fact, without statistics, it's impossible to conceive of development. Without statistics, it's difficult for local authorities to move forward and, above all, to move forward well, i.e. on an objective basis, without trial and error. That's why statistics are essential. This is not a difficult task. At each level of intervention, it's a matter of noting what has been achieved, in terms of human, financial and other resources or tools. At the end of the action, it's a question of indicating what has been achieved in terms of results, measured and recorded in a document. To do this, each action or task must be associated with a sheet to record everything that is or can be converted into a figure.

Statistics can be produced by any technician. Every employee in any public or private service should be concerned with producing statistics for his or her own use and that of other colleagues. Managing the statistics produced is an equally important activity. Statistics are stored and disseminated for use. They must be available, accessible and usable by any user, at any time.

In conclusion, we note that there is a lack of understanding of the realities of local authorities by political, social and economic players and by development partners. As a result, the policies, programs and projects initiated for local authorities are, for the most part, based on an approximate approach. The results obtained are obviously average or weak, while the expectations of the population are numerous and high.

Part 2: ANALYSIS OF DECENTRALIZATION PRACTICE

Chapter 3: Political, social and economic framework

Knowledge of the political, social and economic context in which local development takes place in Côte d'Ivoire is of the utmost importance. It provides information on the general conditions under which political decision-makers, social and economic players are prepared to act, and above all on their level of commitment. The question is whether this context was favorable or unfavorable. To answer this question, we need to carry out a brief analysis of this context, starting with Côte d'Ivoire's accession to independence. Over this sixty-year period, we will highlight the factors that have favoured or accelerated, slowed down or delayed the process of local development in Côte d'Ivoire.

3.1. Assessment of political action

Côte d'Ivoire's political history has had its ups and downs. The farsightedness and vision of President Félix Houphouët-Boigny, and the political stability the country has enjoyed, have been favorable factors in the political management of power.

Côte d'Ivoire is now 40 years into decentralization (1980-2020). The balance sheet in terms of the effective practice of decentralization as a means of solving problems is not very encouraging. Côte d'Ivoire is still grappling with problems of varying nature and complexity. The solutions advocated have not enabled Côte d'Ivoire to take the decisive step towards progress. It can therefore be said that Côte d'Ivoire has yet to prove itself through the decentralization of its power management system.

Gérard Larcher, President of the French Senate, quite rightly said, on the occasion of the Senat forum on local authorities, organized from February 17 to 18, 2020, in Abidjan: "Let'*s not expect decentralization to solve all our problems*" .[1]

It's a warning that needs to be taken very seriously. It means that, in addition to decentralization, we need to take into account other aspects that condition or influence the conduct of local government affairs.

Decentralization in Côte d'Ivoire has been conceived and implemented in a political environment marked by multiple crises over the past thirty years. Today, the overall results of this decentralization leave us wanting more.

All in all, the record of political action is more than mixed. It does not reflect Côte d'Ivoire's ambitions. The idea is that we need to go beyond what has been achieved so far, by being more imaginative, bolder and more responsible in the running of local affairs.

3.2. Social action balance sheet

Social action is measured in terms of the provision of social services according to the expectations or demands of the population. Social demand is determined by the quality of the country's demographic evolution, also known

[1] Fraternité Matin, February 18, 2020, page 3.

as the demographic dividend. This implies the provision of quality social services to the population.

In Côte d'Ivoire, social services were mainly provided by the central government and concessionary companies until the early 1980s. From the mid-1980s onwards, the Ivorian state, under pressure from the Breton Wood institutions through structural adjustment programs (SAPs), disengaged from the direct provision of these services. The post-disengagement state had not been properly prepared. There was therefore a vacuum in the provision of these services. Faced with this situation, public discontent grew.

3.3. Assessment of economic action

Economic action was largely made possible by the availability of land resources, a still cheap labor force, the implementation of five-year development plans, good rainfall and its distribution throughout the year. Agriculture was the main driver of economic activity.

In Côte d'Ivoire, and among international development organizations, there has long been, and still is, talk of falling prices for the main agricultural exports. For decades, the limits of unprocessed exports of the main agricultural crops seemed to have been ignored.

The issue of transforming agricultural produce into semi-finished or finished products was not seen as a priority for the government. Certainly, it was easier for those in power at the time to export raw materials in order to generate immediate, short-term financial resources. No thought was given to identifying and implementing programs or projects to transform raw materials. It was, arguably, a short-sighted policy pursued for years, if not decades. Today, Côte d'Ivoire is paying dearly for the consequences of this policy, based on the all-out export of raw materials.

It should be noted, however, that in the course of Côte d'Ivoire's economic development, some aspects have not been sufficiently mastered. Other aspects have even been ignored. For example, Ivorians were not organized to master the various sectors of economic activity.

The conduct of economic action had not been accompanied by slogans likely to lead Ivorians to take a leading role in local and national economic activity. The absence of mobilizing slogans was seen as one of the most glaring weaknesses in the conduct of economic action. Côte d'Ivoire is paying for this today, with a wait-and-see attitude on the part of citizens who continue to say: "the State must achieve this or that". Ivorians exclude themselves where they should be major players. Functionalism has destroyed their commitment to real, lasting involvement in economic activity. When, at the same time, foreigners have, on the contrary, organized themselves to take control of whole swathes of economic activity. The record of economic action is disappointing, and the prospect of Ivorians taking control of essential areas of economic activity remains dim.

3.4. Threats to local development

The political, wartime and economic crises that marked the life of the nation were particularly detrimental to its progress. The management of the various crises had not been carried out satisfactorily. One of the real threats to local development are the local elections. Côte d'Ivoire always emerges from local elections with after-effects, even dissension, among the population.

After the elections, it's hard to reconcile people. This weakens local authorities as a whole, and deals a heavy blow to the development project proposed to the population. The implementation of this project is therefore mortgaged. The day when local election candidates come forward with the intention of genuinely working for the progress of the community, and therefore in the interests of the people, we will no longer see such hateful political rhetoric, based on lies and cheating.

3.5. Resources mobilized for local development

Human resources are examined in terms of their relevance to local development objectives. The training and qualification of local authority staff is a strategic issue. The professional experience of staff, their level of commitment to the job, the allocation of roles and responsibilities, and their motivation are all factors that determine the performance of local authorities.

An examination of the financial resources needed by local authorities for general operations and productive investment shows that they have hardly lived up to expectations. In reality, local authorities have never had the financial resources to work. The small budgets of local authorities have made it possible to carry out small projects. This can be interpreted as a lack of ambition on the part of the public authorities. This situation has not yet led to adequate solutions to the problems facing these communities.

Technical and logistical resources are examined in terms of working materials, office equipment, vehicles and other work equipment. Today, local authorities are still poorly equipped. There is also a lack of rigor in the management of working materials granted to local authorities, including motorized vehicles. Local authority vehicles are regularly used to transport goods and property belonging to private individuals. In fact, there has been a great deal of laxity in the management of local authority assets.

3.6. Managing economic partnerships

The development of socio-economic partnerships is an increasingly recommended option for local authorities. These partnerships enable them to open up to the outside world. The content of these partnerships must be sufficiently clear for all stakeholders to better understand their implications.

What's more, the management of these economic partnerships is not sufficiently optimal. As a result, it does not always meet the requirements of the partners. The purpose of an economic partnership is to have a real impact on local development initiatives, and consequently on the lives of partner communities.

The political, social and economic framework in which local authorities have evolved to date has not been conducive to their development. It has not enabled programs and projects to be carried out in line with Côte d'Ivoire's stated ambitions. It has not provided the human and financial resources needed to carry out development initiatives.

Chapter 4: Decentralization and Abidjan's macrocephaly

Côte d'Ivoire's urban fabric is characterized by the dominance of Abidjan. A macro-cephalic city, Abidjan is the pride of its leaders and inhabitants alike. But, at the same time, it remains a "perfect" concentration of modern-day urban problems.

For our country, we maintain that Abidjan's predominant position in the urban fabric is a handicap to the balance of the national territory. Abidjan's position is identical to that of the capitals of the Black African countries colonized by France. This chapter presents the essential aspects of this macrocephaly, and outlines a number of possible solutions, with a view to achieving more balanced development throughout Côte d'Ivoire.

4.1. Background: "Paris and the French desert

The question is whether Abidjan's macrocephalic position is an asset or a handicap for Côte d'Ivoire's development. France experienced this situation with the city of Paris. The "City of Light" concentrated everything, crushed everything, crushed everything in France. Paris et le désert français" is the title of a book written by Jean-François Gravier and published by Portulan in 1947. There was no desert in France. The author of the book, which was considered the bible of decentralization, had wanted, with this colorful title, to say that in France, there is Paris. And apart from Paris, there was no other city comparable to it. The rest of France had no cities to counterbalance the weight of Paris. In the aftermath of the Second World War, the young geographer, only 32 years old, was advocating solutions to alleviate Paris's macrocephaly.

Major reforms, inspired by this book, had been undertaken in regional development planning. In the decades that followed, these reforms led to the creation or reinforcement of metropolitan areas in several regions, thereby balancing the development of the French territory.

Abidjan's current situation is identical to that of Paris in 1947. It calls our attention to the absolute necessity of undertaking, here too, reforms to counterbalance, over the next 30-40 years, Abidjan's excessive weight on the national urban scene.

4.2 Characteristics of Abidjan macrocephaly

On the administrative and political front

Abidjan is home to the headquarters of administrative services, the Republic's major institutions, diplomatic representations (embassies, consulates) and international development cooperation organizations. Abidjan remains the main center for administrative and political decision-making. Yamoussoukro, designated the administrative and political capital since 1983, is still struggling to assert itself.

Demographics

Abidjan, the economic capital of Côte d'Ivoire, is home to a population of 4,395,243 or 40% (2014 RGHP) of the country's urban population. According to the same census, the population of Bouaké and Daloa, the country's second and third-largest cities, is 536,719 and 245,360 respectively. The population of the two cities is 782,079, or just 17.79% of that of Abidjan.

On the academic, scientific and technological front

Abidjan alone concentrates almost all of the country's academic, scientific and technological potential. Abidjan's overwhelming domination of the country's scientific and technological expertise is symbolized by the presence of two public universities (the largest in the country). In addition to these public universities, there are a multitude of private universities, faculties and grandes écoles. Abidjan's scientific and technological clout is also reflected in its technical design offices, consultancies, law firms, accounting firms and legal practices.

On the economic front

Abidjan has four (4) industrial zones, located respectively in Vridi, Koumassi, Yopougon and Akoupe-Zeudji (PK 28 of the Autoroute du Nord). The city of Abidjan is home to the bulk of the country's industrial production, processing and marketing fabric. It is, without doubt, the main center of economic activity.

On the health front

Abidjan is home to four world-class University Hospitals (CHU), in Cocody, Treichville, Yopougon and Angré respectively. It is also home to dozens of health clinics and polyclinics, some of which are world-class.

In terms of services

Abidjan is home to most of the services provided to the population of Côte d'Ivoire (it doesn't have everything that cities of its rank around the world have). It is home to banks, insurance companies, shops, supermarkets, top restaurants and more.

On the tourism and hotel front

Abidjan boasts a dozen world-class hotels, enabling it to host international meetings and conferences with hundreds or thousands of participants.

Leisure and culture

Abidjan is home to the Palace of Culture, libraries, cinemas and concert halls. Abidjan is home to musicians and musical groups of national and international renown.

On the airport front

Abidjan has the largest airport in Côte d'Ivoire (and even in the UEMOA region). Abidjan Port-Bouet airport records some forty flights and landings daily, compared with one or two flights or landings for San Pedro.

In terms of road infrastructure investment

Abidjan currently accounts for the bulk of investment in road infrastructure. These include bridges and interchanges.

4.3. The need for reform

4.3.1. Reforms to balance Abidjan

Reforms are needed to plan the development of this large geographical area we call the **"Grande Couronne d'Abidjan"**. The "Grande Couronne" goes, of course, beyond Abidjan. It extends to around two hundred kilometers around Abidjan. By 2030-2040, balanced development of this spatial unit will help relieve congestion in the current Abidjan conurbation. The "Grande Couronne d'Abidjan" will be developed around the following three (3) poles:

- **POLE 1 (West)**
  Dabou, Grand-Lahou, Fresco, Tiassalé, Divo.

- **POLE 2 (East)**
  Grand-Bassam, Bonoua, Adiaké, Aboisso, Tiapoum, Noé.

- **POLE 3 (North)**
  Anyama, Agboville, Adzopé, Bongouanou and Abengourou.

4.3.2. Reforms to promote regional clusters

The current functioning of the Abidjan conurbation is not optimal. This situation suggests that, even with the most optimistic economic assumptions, the functioning of Abidjan in 2030-2040 will continue to pose serious problems for our nation. Solving Abidjan's problems in this timeframe must now be seen from the perspective of identifying and promoting regional hubs of sustainable human, economic, scientific and technological development.

The hubs are arranged in a hierarchy, from the largest to the smallest. Investments should be judiciously distributed according to the size of each hub, its vocation and the degree of deterioration of its infrastructure.

4.3.3. Suggested spatial organization reforms

Côte d'Ivoire's entire national territory can be swarmed by large spatial clusters, with a social, economic, scientific or technological vocation. By way of example, the following development clusters or groupings of clusters can be defined:

Table: Côte d'Ivoire's development hubs

N°	Poles	Main city
1	Odienné Boundiali	Odienne
2	Touba Seguela Mankono	Seguela
3	Korhogo Ferké Kong	Korhogo

4	Man Danané Guiglo Duekoue	Man
5	Daloa Issia Vavoua Zuenoula	Daloa
6	San Pedro Sassandra Grand-Bereby Tabou Grabo	San Pedro
7	Gagnoa Soubré Lakota Oumé	Gagnoa
8	Yamoussoukro Bouaflé Toumodi Dimbokro Tiebissou Didievi	Yamoussoukro
9	Bouaké Béoumi Sakassou	Bouaké
10	Daoukro Mbahiakro Bocanda Prikro	Daoukro
11	Katiola Satama-Sokoura Dabakala	Katiola
12	Bondoukou Tanda Bouna	Bondoukou

In all, in addition to the greater Abidjan area, Côte d'Ivoire will be divided into twelve (12) development poles or groups of poles. The development, equipment and rational management of these poles will considerably reduce not only regional disparities, but above all Abidjan's weight in the national accounts. This suggestion deserves dispassionate consideration.

4.4. Potential of Abidjan's inner suburbs

Identifying all the potential of the Greater Abidjan area (natural resources and environment, human resources, economic resources) is of prime importance. A detailed, objective analysis of these potentialities will enable a relatively precise diagnosis to be made. On the basis of this diagnosis, strategic development plans and sectoral development projects will be defined or identified, programmed and implemented according to an established schedule. Data collection and analysis for the Greater Abidjan area will take two years.

Regional planning for the Greater Abidjan area

The Grande Couronne area of Abidjan needs to be properly developed and equipped to enable it to play its role as a driving force for development. To achieve this, the following reference documents and instruments need to be put in place:

Regional documents and instruments

- Schéma Régional d'Aménagement et de Développement de la Grande Couronne d'Abidjan ;
- The Greater Abidjan Development Plan;
- Observatoire Régional d'Aménagement de la Grande Couronne d'Abidjan ;
- Economic Accounts for the Greater Abidjan Area.
- L'Atlas de la Grande Couronne d'Abidjan.

Local documents and tools

- Each entity's Urban Master Plan (SDU);

- Each entity's Urban Master Plan (UMP);
- The Detailed Urban Plan (PUD) for each entity;
- The subdivision plan for each entity;
- Each entity's Strategic Development Plan.

Financial instruments

- Fonds d'Appui à l'Aménagement et au Développement de la Grande Couronne d'Abidjan;
- Financial contributions to sub-regional and regional integration ;
- Financial contributions from development partners.

Coordination and monitoring-evaluation arrangements

- Direction Générale de l'Aménagement de la Grande Couronne d'Abidjan ;
- Les Conseillers d'Aménagement de la Grande Couronne d'Abidjan.

4.5. A tentative typology of possible reforms

The implementation of development programs and projects in the Greater Abidjan area must be based on a diversified set of reforms. The reforms to be implemented concern all sectors of activity. The following typology is proposed.

Defining an attractive investment policy

- Demonstrate a firm commitment to attracting investment to the Greater Abidjan area.

Administrative procedure for investment

- Facilitation of administrative procedures to speed up the processing of applications submitted by economic players;
- Review of various investment procedures.

Optimization of various sector development codes

- Investment code ;
- Forestry code ;
- Rural and urban land code;
- Mining investment code ;
- Agricultural investment code ;
- Tourism and hotel investment code;
- Code de la construction et de l'habitat.

4.6. Reform benefits

Land and housing benefits

- Creation of an urban land development organization;

- Provision of serviced and equipped land for economic operators;
- 50% reduction in the purchase price of developed and equipped industrial land in the Greater Paris area;
- Free registration of the business or industrial operator's land ;

Economic and customs advantages

- Offer facilities for setting up economic activities;
- Free registration for the structure ;
- Granting of economic and customs facilities on imports of production machinery and tools.

Financial benefits

- Creation of a bank to finance the development of the Greater Abidjan area;
- Financial assistance for the relocation and installation of an economic, scientific or technological activity in the greater Abidjan area;
- Financial assistance for the initial establishment of an economic, scientific or technological activity in the inner suburbs;
- 50% rebate on electricity, water, telephone and Internet connection charges.

Tax advantages

- Tax benefits for companies wishing to relocate to the Greater Paris area for three years;
- 50% tax rebate on BIC for all companies wishing to relocate to the inner suburbs, for a period of three years.

Professional training benefits

- Exemption from payment of vocational training taxes for three years;
- Training and capacity-building for employees of companies based in the Greater Paris area.

Legal advantages

- Free legal services and assistance for any company wishing to set up in the area;
- Reinforcement of legal texts on investment protection in the Grande Couronne.

4.7. Development of economic, social and environmental education-training

Road infrastructure

- Densification or reinforcement of the Grande Couronne road and motorway network through the opening of new roads;
- Improvements to the existing road network in the Greater Paris area.

Drinking water supply

- Investment in drinking water supply in the Greater Paris area;
- Reinforcement of the existing water supply network through the construction of new water towers.

Electricity supply

- Use of conventional energy ;
- Promoting renewable energies.

Telecommunications and the digital economy

- Improvements to conventional telecommunications ;
- Promotion of digital communications.

Management of natural resources and the environment

- Programs and projects for the rational management of natural resources ;
- Rational environmental management programs and projects ;
- Programs and projects to combat climate change.

Digital economy and business intelligence

- Promoting the digital economy ;
- Promoting business intelligence.

Education and training

- Creation of high-level universities and technical schools for the initial and vocational training of workers for the local economy;
- Matching training programs to the needs of the local economy ;
- Promotion of work-study programs with companies in the Greater Paris region (zero unemployment for graduates).

Housing and home ownership

- Improving conditions for habitat creation;
- Eliminate all forms of substandard and precarious housing by promoting housing for the underprivileged (adapted financing methods, mutualist forms of housing production).

Job creation investment program

- Creation of entrepreneurship development centers in the Greater Montreal area;
- Creation of study, assistance and assessment centers for businesses;
- Creation of an investment fund for job creation and development;
- Creation of a fund to support youth entrepreneurship;
- Expansion and modernization of businesses created and managed by young people and women.

Circular economy promotion program

- Responsible consumption of natural resources in the Greater Abidjan Area ;

- Responsible consumption of raw materials in the Greater Abidjan area ;
- Production and rational management of solid waste in the Greater Abidjan Area;
- Management of opportunities linked to the circular economy for job creation in Abidjan's Grande Couronne ;

Promoting the tourism industry and quality of life

- Tourism development plan for the region ;
- Quality of life in the region.

Promoting business information

- Data collection and analysis (economic data bank)
- Dissemination of economic information.

This chapter calls on government authorities to examine the future of the city of Abidjan over the next 30-40 years. This future can only be envisaged and built with the recommended development hubs or clusters.

Our most fervent wish is that, on the basis of this reflection, the public authorities will take the boldest initiatives. One such initiative, for example, would be to set up a program of sectoral studies to determine the feasibility of the Grande Couronne d'Abidjan. The strategic development plan resulting from these studies would detail the reforms needed to achieve the expected results.

Chapter 5: Responsibilities and powers transferred

Côte d'Ivoire has transferred responsibilities and powers to local authorities. Government authorities have acted to increase the powers of these communities. Naturally, the latter applauded. They were disillusioned, however, because overnight they found themselves with responsibilities or powers that they were unable to exercise effectively, due to the absence or inadequacy of human, legal and financial resources to support the transfer. Legally speaking, the various decrees implementing this law have not been issued, making it virtually impossible to apply.

5.1. Responsibilities or powers transferred

LOI N° 2003-208 DU 07 JUILLET 2003 PORTANT TRANSFERT ET REPARTITION DE COMPETENCES DE L'ETAT AUX COLLECTIVITES TERRITORIALES

The responsibilities or powers transferred by the State to Ivorian local authorities are :

- ➢ regional planning ;
- ➢ development planning ;
- ➢ urban planning and housing ;
- ➢ roads and networks;
- ➢ transport ;
- ➢ health, hygiene and quality;
- ➢ environmental protection and natural resource management ;
- ➢ safety and civil protection ;
- ➢ education, scientific research and vocational and technical training;
- ➢ social, cultural and human development activities;
- ➢ sports and leisure ;
- ➢ promoting economic development and employment ;
- ➢ tourism promotion ;
- ➢ communication ;
- ➢ hydraulics, sanitation and electrification;
- ➢ promotion of the family, youth, women, children, the disabled and the elderly.

5.2. Authority to manage transferred powers

Careful and objective observation of the responsibilities or competencies in this list, in view of the current situation of local authorities in Côte d'Ivoire, leads us to make the following comments regarding their exercise:

- From our point of view, the number of powers (16) transferred to local authorities is excessive;
- The decision to transfer these skills to local authorities is clearly not in line with what exists or what is real;

- As things stand at present, it is absolutely impossible for the country's local authorities to take these skills into account, due to the absence or inadequacy of human resources, both quantitatively and qualitatively, and the low "technical absorption" capacity of the authorities (inadequacy or absence of knowledge or know-how, or know-how to do things within them);
- The absence of implementing decrees has meant that the roles and responsibilities of local players have not been clarified to ensure that skills are truly and effectively taken into account.

The transfer of responsibilities and competencies is unrealistic. This transfer is also cumbersome, while local authorities, in their current state of development, have little capacity to absorb or manage these competencies.

5.3. Transferring responsibilities or competencies - and then what?

Skill transfer is an essential question. It deserves to be asked. An analysis of all the official comments made on local authorities in Côte d'Ivoire in recent years generally reveals the inadequacy of their structural and strategic organization. One example is the statement made by Ivorian government spokesman Bruno Nabagné Koné: "*Move more quickly towards an organization that promotes efficiency and development. The region and the commune are the two most appropriate types of local authority to promote local development and ensure the full involvement of the population in the management of their affairs*" . [2]

What does the Ivorian Constitution say about the financing of local authorities? It is not silent on the matter. Thus, we note that Article 174 of the Constitution of October 30, 2016 stipulates: "*Any transfer of competencies between the State and territorial authorities shall be accompanied by the allocation of resources equivalent to those that were devoted to their exercise*"[3] . The law says it all. But in reality, the State has almost never met this condition. Everything suggests that, for the Ivorian State, simply transferring responsibilities or competences is enough for everything to fall into place. Transferring responsibilities or powers to local authorities is only one step. The power to actually exercise these responsibilities or competences is the next step. But this power does not exist. Local elected representatives cannot freely decide what needs to be done, because they lack the human and financial resources to carry out their mandate. Even when local authority budgets are adopted or approved each year, they are not implemented one hundred percent of the time. Local councillors are crying out in dismay at the drastic drop in financial resources granted by the State.

All in all, we find ourselves in a context where the Ivorian State does not seem to have truly prepared itself before transferring so many responsibilities or competencies to local authorities. They do not have the power to mobilize

---

[2] Statement made at the end of the Council of Ministers meeting held on September 28, 2011 in Yamoussoukro.

[3] Dagobert Banzio (SG of Ardci), " L'Etat nous a transféré 16 compétences sans que les moyens ne suivent ", in Fraternité Matin of February 28, 2017, page X

human and financial resources. They have yet to adopt approaches capable of enabling them to emerge from the situation thus created.

5.4. Understanding how communities work

5.4.1. Bodies of Ivorian local authorities

For their operation, Ivorian local authorities have the following three (3) bodies: the deliberative body, the executive body and the executive support body.

1°) **The deliberative body** is made up of Councillors. These include the Regional Council, the District Council and the Municipal Council.

2°) **The executive body** is the person who chairs the Council, executes its decisions and manages the day-to-day administration of the community. These include the President of the Regional Council, the District Governor and the Mayor of the Commune.

3°) **The executive's support body** is the restricted college made up of the executive and his deputies (4 to 6). Thus, there is the Regional Council Office, the District Council Office and the Municipality.

In addition to these 3 types of bodies, some local authorities also have consultative bodies, which must be consulted beforehand on issues falling within the remit of the local authority (plans, programs, budgets, etc.) and are made up of representatives of civil society: the Regional Economic and Social Committee for the Region, the District Consultative Committee for the District. We note that the Town Hall does not have a consultative body. And yet, given the scale of the socio-economic problems facing the communes, the creation of a consultative body would not be superfluous.

5.4.2. Community operating procedures

The executive authorities of local authorities convene the Bureaux or Municipalities at the headquarters of the decentralized entity at least once a month, and whenever required to settle matters falling within their remit. However, the supervisory authority may authorize Bureaux or Municipality meetings to be held in places other than the said head office and located within the perimeter of the local authority.

Council Boards and Municipalities can only validly deliberate on the agenda if at least half their members are present. If the quorum is not reached after the first meeting has been duly convened, the decision taken after the second meeting has been convened at least eight days later is valid, regardless of the number of members present.

However, in times of war or calamity, the text specifies, the Boards of Councils and Municipalities deliberate validly regardless of the number of members present. Decisions of the Boards and Municipalities are taken by relative majority. In the event of a tie, the vote of the local authority's executive authority is decisive. Meetings of the Boards and Municipalities are not open to the public. The Boards and Municipalities may invite to attend their meetings, in an advisory capacity, any persons whose presence they deem useful.

The minutes of Council and Municipal Board meetings must mention the identity of absentees and the decision taken as to the legitimacy or otherwise of the reasons for absence. Any unexcused absence is deemed to be illegitimate.

The minutes of the meetings of the Offices and Municipalities are communicated to the Councils at their next meeting.

In public ceremonies, and whenever the exercise of their functions makes it necessary, members of Bureaux and Municipalities wear a sash tied around their waist as a distinctive sign of their office. This scarf, in the national colors, is made up of three thirty-three millimeter bands with gold bangs and tassels at the ends for authorities vested with executive power in local authorities, and silver fringes and tassels for other members of Bureaux and Municipalities.

5.5. What does Côte d'Ivoire really want?

Careful and objective observation of the list of transferred responsibilities or competencies, given the current situation of local authorities in Côte d'Ivoire, leads us to ask this question. The answer to this question should enlighten us. Côte d'Ivoire wants to put in place the elements for locally-based development. Everyone agrees on this overall objective. At the same time, we say that Côte d'Ivoire has not, to date, given itself the means to achieve it. When we talk about means, we're referring to human and financial resources, legal elements and the know-how of elected representatives. To date, these resources are not sufficient to meet people's expectations.

It's not too late for Côte d'Ivoire to review the whole issue of decentralization. It must do so objectively, without political calculation. The concern of the central authorities is to enable local authorities to really work. By working effectively, they lighten the workload of central government. That's what decentralization is all about.

5.6. What are the prospects for managing transferred skills?

Given this situation, what should really be done? In other words, what are the prospects for Côte d'Ivoire? For our part, we put forward the following two hypotheses:

- ✓ First hypothesis: The Ivory Coast government maintains the powers transferred to the local authorities as they stand. It then takes the time to prepare the various implementing decrees for each of the transferred powers. Then, it agrees to provide local authorities with adequate human, financial and logistical resources, as well as the necessary knowledge and know-how, to enable them to function properly and make appropriate investments.

- ✓ Second hypothesis: The Côte d'Ivoire government decides to transfer responsibilities or competencies to local authorities in stages, gradually. In this case, the State proposes to prepare the local authorities to get used to the process over the years. In particular, over a period of three or five years, it can help them acquire the knowledge and know-how they need to manage their responsibilities in the best possible way. This learning and capacity-building period would enable local authorities to move up the ladder in the management of transferred responsibilities or competencies, according to a consensual timetable.

5.7. Effects of multiple functions

Cumulative office-holding means that, in addition to holding the position of chief magistrate of a locality, the elected official also holds another public, parapublic or private office. For example, a man or woman may be elected mayor of a commune and hold a position as director of a public or parapublic administration, or as minister or vice-minister.

Holding multiple offices has a direct and negative impact on the running of the community's activities. For example, an elected official who holds more than one position may not have enough time to carry out his or her duties effectively. Of course, they can rely on their ability to organize their activities. With today's digital resources, it is indeed possible to save time. But, generally speaking, the elected official is in a chase to make up for lost time, to put things back in order. Generally speaking, they are obliged to complete certain tasks in a hurry, which is a source of error.

On the other hand, holding multiple positions means holding multiple financial resources. Each position entitles the holder to a salary or allowance. However, the pursuit of financial gain through multiple offices can be detrimental to an elected official's ability to carry out his or her duties.

In addition, in the eyes of local public opinion, the holding of a number of offices has given rise to sometimes acerbic comments about or on behalf of the elected representatives concerned.

Finally, it would be wise for local councillors who hold multiple offices to be honest enough to relinquish some of them, all the more so as the management of a local authority, for greater efficiency, requires the close involvement of the councillor, who must oversee all aspects of the institution's life.

Faced with the many disadvantages of holding multiple offices in the management of local authorities, it's up to the public authorities to assume their responsibilities and put an end to this practice. Such a decision would be salutary, not only for the elected representatives themselves, but also for their communities and their populations. Today, Côte d'Ivoire has the potential for high-quality human resources in various fields of human activity. It's not fair to see that, while some players are accumulating functions, others, more and more numerous, are without any real occupation. We need to make a reasoned decision.

Chapter 6: Status of decentralization in Côte d'Ivoire

The current state of decentralization in Côte d'Ivoire is nothing to crow about. Some observers of decentralization point to undeniable results, particularly in terms of improving the level of infrastructure facilities available to local authorities. Others, on the contrary, point to a certain sluggishness in Côte d'Ivoire's march towards progress. Other observers, more severe in their judgement, feel that there have been few results compared with the hopes raised when municipal development was introduced in 1980. In the face of these divergent opinions, it is worth taking an objective look at the situation of local authorities to realize that it is not as glowing as one might think. To illustrate this, we need to highlight a few aspects of the national decentralization system.

6.1. Decentralization policy objectives

In Côte d'Ivoire, since 1980, decentralization has been seen as a major thrust of the policy of successive governments. None of them has deviated from the path of decentralization. On the contrary, each government has tried to add its own personal touch, according to its vision of local development.

Emile Constant Bombet, Minister of the Interior and Decentralization under President Henri Konan Bédié, gave the following definition of decentralization: "*Decentralization is defined as a modification of State power, as a technical process that grants a greater or lesser degree of freedom to local authorities to regulate their own affairs. It therefore consists in conferring decision-making powers on bodies of their own, distinct from those of the State. It is a system in which local decision-making plays a major role, without infringing on the national interest, insofar as the ultimate responsibility for social well-being lies with the State. Finally, it is set up to create the objective conditions for balanced, self-sustaining development*" .[4]

Minister Constant Bombet goes on to say that: "*The aim of decentralization is to encourage the involvement of local populations in matters that concern them first and foremost, as well as the animation of local life, with a view to promoting and facilitating development. Decentralization is conceived as the most effective means of achieving a coherent, and therefore sustainable, choice of investments to be made*".

Decentralization is taking place in a national context characterized by profound changes in various sectors of human activity. We are witnessing political, social, cultural, economic and environmental changes throughout the country. The national decentralization movement is irreversible. It is being implemented with these changes in mind.

6.2. Diagnosis of the socio-political context

[4] **BOMBET, Emile Constant**. Introductory presentation of the government's decentralization and regional planning program, at the Donors' Round Table on "*Decentralization and Regional Planning*", held in Yamoussoukro, May 12-14, 1997.

6.2.1. On the political front

In terms of decentralization policy, it is worth recalling that there was a political will to opt for this form of governance of the national territory. The intention to build the country's progress from the ground up was real. However, it should be pointed out that the translation of this will into development programs and projects has not been fully effective. A number of factors may explain this situation. First and foremost, there has undoubtedly been the policy of transferring responsibilities and competencies from central government to local authorities. This policy has not been sufficiently understood or based on the realities of the country. In other words, there was an "ambitious" transfer which, unfortunately, was not based on the capacities of local authorities to manage these competencies. Ultimately, the transfer of responsibilities or competencies is a process that has not really been completed. It appears, in a way, like an unfinished symphony.

6.2.2. Administrative environment

The current administrative status of local authority representatives raises a number of concerns. It does not enable them to carry out their mandate effectively. In material terms, for example, this status offers municipal councillors allowances that are deemed derisory in relation to the work they are required to do. In any case, this status is not motivating for elected municipal officials. Observers of local authorities rightly point to absenteeism among elected representatives. This absenteeism is perceived as the result of their inability to meet the many needs of the population. Mayors seem to be running away from the people to whom they have made so many promises. This is a source of great frustration for the local population, who want to see them, meet them and hear what they have to say about the issues that concern them.

Local authority workers are no better off. Their status also poses problems. The status of local civil servants, the freeze on promotions and salary arrears are recurring issues. Their status does not guarantee them attractive rights and benefits. These workers complain of the absence of an attractive career profile. The question of the status of local authority workers raises that of their responsibility, the sustainability of their employment and their professional mobility. The level of their remuneration does not motivate them to increase their individual and collective performance.

6.3. Diagnosis of urban services management

Since the economic crisis of the mid-80s, new social infrastructures have not seen the light of day and, generally speaking, existing infrastructures have seen their physical condition deteriorate. This situation has worsened with the decline in state resources and the political crises the country has experienced. Over and above all that can be said about the failure to provide basic social services, we note that this is a problem of governance.

6.3.1. Drinking water supply

The supply of drinking water to local communities remains one of Côte d'Ivoire's greatest challenges. In many parts of the country, women and girls continue to obtain their water from rivers, marigots and ponds, with the attendant health risks.

"I *had an appointment this Tuesday, March 24, 2020, with Dr Koutoua Amon Jean-Pierre, Head of Technical Services at Issia Town Hall. I arrived at his office at ten in the morning from the town of Issia. From his office, I called him to say hello. He replied that he was still at home, and that he was about to go to his office.*

His secretary then informs me that his boss always arrives at the office very early. *But," she hurried to inform me, "for the last four days, the water supply has been interrupted in the town of Issia. As a result, the guards take two cans of water to the Grotte, located in another district, to bring it to the head of technical services for washing. After that, he can go to the office*".

This situation is not specific to the town of Issia, which has a population of 86,000. It's the situation in dozens of other towns, large and small, including Daloa, Côte d'Ivoire's third-largest city. It's easy to see why people in some of the country's villages still get their water from the marigot. What government hasn't made promises to the people of Côte d'Ivoire's towns and villages during the local election campaign? The result is that, in 2020, drinking water remains a rare commodity in our local communities.

6.3.2. Provision of school services

In 2020, many pre-schools and primary schools are in an astonishingly precarious physical state. There are, in fact, nursery and elementary school built of rammed earth, schools with straw or papot roofs, often without walls. Generally speaking, in Côte d'Ivoire, many existing school infrastructures are in a disastrous state due to lack of maintenance. Many secondary schools are also in a deplorable physical state. For these establishments, the walls are cracked, allowing rainwater to seep through, with inevitably high levels of humidity in the classrooms. Ceilings hang like a sword of Damocles over the heads of teachers and pupils. There are also defective or obsolete classroom lighting systems, damaged or painted walls, tattered woodwork (doors, windows and cupboards) and plumbing systems. Throughout the country, educational establishments, particularly nursery, primary and secondary schools, are in need of renovation.

Today, there is a shortage of schools in local authorities, with the corollary of overcrowding in existing classrooms, which is contrary to the provision of good teaching. There is also a crying shortage of teachers, key players in the system.

Côte d'Ivoire is forced to plug the gaps with cut-price teacher recruitment programs. Hastily recruited and trained, and recklessly thrown into the already moribund education system, most of the new teachers recruited through these programs have no vocation to embrace and practice the teaching profession. All of which goes to show the sorry state of education in Côte d'Ivoire.

6.3.3. *Access to* healthcare *services*

Local authorities provide health services and care for their populations. The problem is how people access these services and care. How do people care for themselves in local communities? Because of poverty, many people are forced to turn to traditional medicine for their health problems. During the years of turmoil, the public authorities, whether central or local, have not invested in the construction of local health facilities or in the training of health personnel. Access to health care and services, particularly at local or community level, has become a real problem for the population.

6.3.4. *Waste and sanitation in towns and villages*

Waste and garbage management in the country's urban centers and villages is not optimal. There's a lot of talk (including childish slogans). In the end, the results are not conclusive. They fall far short of expectations. The omnipresence of uncollected solid waste in the towns and villages of Côte d'Ivoire is a national disgrace. In reality, Côte d'Ivoire has not been able to find the approaches and strategies needed to respond sustainably to the waste challenge. Today, waste collection does not, paradoxically as it may seem, involve local authorities. They are not responsible for anything when it comes to waste and refuse management.

6.3.5. *Community electrification*

The country's electrification is only halfway complete. In many cities, there are neighborhoods without lighting. Many villages are still in darkness. Promises made by successive governments have not been kept. This situation makes it impossible to lay the foundations for local development, as many initiatives are stifled by the absence of electricity. Renewable energy sources have not been promoted, owing to a lack of clear political will and know-how in the field.

6.4. Diagnosis of the economic context

6.4.1. Real potential but low impact

Côte d'Ivoire's economic potential is very great, compared to many other African countries. No one can doubt this potential. Until the mid-1980s, Côte d'Ivoire had invested quite heavily in economic infrastructure. Between 1995 and 2010, there was virtually no investment in the country's economic infrastructure. The multiple political crises of this period had very negative effects on the existing economic infrastructure and on the country's economic progress. This was reflected, in particular, in the absence of major public investment programs and the physical deterioration of infrastructures and services built during the boom years. In short, Côte d'Ivoire's economic potential has not really been harnessed to improve people's living conditions. From 2011 to 2020, there has been renewed investment in economic infrastructure. But there is still work to be done throughout the country.

6.4.2. Rampant poverty

Today, poverty is omnipresent in the urban centers and villages of Côte d'Ivoire. A large proportion of the working population is idle. Men and women in full possession of their physical abilities, sitting in the shade of trees, in cabarets or coffee or tea kiosks, spending their time talking about subjects of no interest to them. Idleness has led these men and women to criticize everything. It produces hatred, jealousy and violence.

6.4.3. Community social funds

Some local authorities have set up so-called social funds. These funds have been set up to enable people, particularly young people and women, to engage in income-generating activities. The creation of these funds meets a demand. However, the results achieved remain somewhat perplexing. In a way, social funds are a victim of the overly social nature of their creation and management. Indeed, since they are social in nature, there was no obligation on the part of beneficiaries to achieve results. The promoters of social funds did not think about their sustainability. What's more, there was no guarantee that the funds would be granted, which meant they could not expect a certain level of efficiency. The choice of beneficiaries (and projects) is made on a less-than-objective basis in the communities. Furthermore, fund beneficiaries have not received even cursory training in, for example, simplified accounting. Last but not least, beneficiaries have not received any technical support, due to a lack of local capacity to provide it. As a result, not every project selected by the local authorities was carried out effectively. The results obtained are disappointing in most of the country's communities.

6.5. Organizational and communication diagnosis

6.5.1. Rational organization

To take root and achieve long-term success, local development must be based on a rational, methodical, rigorous and controlled organization of activities. However, an objective, independent analysis of local development conditions in Côte d'Ivoire reveals the political instability and even disorder that has prevailed. This has prevented Côte d'Ivoire from building on the momentum of the 60s and 70s. In fact, it came to a standstill. It was as if, during this period of political and social instability, there was no organization, no direction, no compass to guide the country's progress.

6.5.2. Mobilizing workers

The country's political instability, and the economic and social uncertainty that went with it, made it impossible to release the necessary creative energy among its citizens. It was not possible to motivate workers, even though salaries were not regularly paid in some communities. Local authority workers paid the highest price for the country's political crisis. There was a period when the productivity of local authority workers declined, with multiple strikes.

6.5.3. Communication within communities

Communication is essential to the organization and operation of a local authority. It is carried out at at least three levels.

First and foremost, it takes place internally, i.e. within the local authority itself. A good internal communications strategy can mobilize all the players in the local authority to understand their role first and then take action.

Secondly, communication takes place between communities in the same region to create a bond of solidarity. In Côte d'Ivoire, communication between communities in the same region is virtually non-existent. Each community goes its own way, without informing the next. The purpose of inter-community communication is to enable one community to find out what its neighbor is doing. It's a way of sharing experience.

Thirdly, communication takes place between the local authority and the outside world. Here too, local authorities in Côte d'Ivoire communicate with the outside world, without informing those in the region.

Communication at these three levels enables the local authority to reach out to all sections of the population, and to raise awareness among a variety of players, both local and external, with the possibility of mobilizing partners. Last but not least, communication means preparing messages and sending them to target audiences.

6.6. Diagnosis of partnership coordination and management

6.6.1. Coordination of activities

The organization of local authorities in Côte d'Ivoire places the General Secretariat in a position of central administrative coordination of activities initiated and carried out by other departments.

6.6.2. Partnership management

The willingness of elected representatives to develop partnerships is commendable. It should be encouraged. But willingness is not enough. The development and management of partnerships by local authorities in Côte d'Ivoire has its weaknesses. In particular, there is a lack of know-how, inefficiency and, above all, slowness in carrying out the actions included in the program. In any case, there is a clear lack of expertise in partnership management.

6.7. Performance management diagnosis

Local authorities have overall objectives to achieve, through the development plans they adopt. The achievement of development objectives is based on indicators that serve as a scorecard. The current level of performance of local authorities in Côte d'Ivoire is judged to be average or even poor. Assessing the performance of local authorities is an exercise of strategic importance. This assessment must be carried out objectively by an independent source.

6.8. Diagnosis of local authority financial management

Some of the weaknesses noted in the management of local authorities in Côte d'Ivoire concern finance. These weaknesses seem, in part, to be the consequence of the current status of elected representatives. Four (4) weaknesses are generally identified when assessing the actions of elected officials. These are irregularities that generally concern the following cases:

- Over-invoicing for acquisitions;
- Embezzlement of funds and materials;
- Acts of de facto management ;
- Corruption.

The information at our disposal enables us to give an overview of the practice of overbilling in local authorities. For the purposes of this work, we have selected two overbilling practices: i) the purchase of a bus; ii) the construction of a social or economic facility.

Buying a coach

The local authority official responsible for conducting the coach acquisition process invites or visits the dealer's representative to propose the deal. The dealer's representative indicates the actual selling price of the coach. He says that the coach costs, for example, 60 million CFA francs on the local market. The local authority official suggests to the dealer's representative: "*You can increase the invoice to 80 or 100 million*". The concessionaire's representative draws up the proforma invoice at this price. At the time of payment, the local authority receives, in the form of a "rebate", the sum of 20 or 40 million CFA francs, part of which is certainly paid to the concessionaire's agent. In this way, the local authority is paying more for the bus, because it has clearly been overcharged. This practice is common for the acquisition of goods and equipment.

Equipment construction

The example concerns the construction of a communal or neighborhood market. The cost of building the market is estimated, for example, at 250 million CFA francs. Generally speaking, the market built does not reflect the cost indicated. Naturally, theft is suspected. The public authorities can commission a counter-assessment of the construction in question. To this end, a building engineer or even a quantity surveyor can determine the real cost of the work. The latter proceeds by corps d'Etat. He takes body of state after body of state. He determines the quantity of materials and equipment used at the time of construction. He uses or exploits the price list in force at the time. At the end of his appraisal, he can accurately state the actual construction cost of the contract in question. His expertise will not be contested.

That's why we believe that if we were to assess all the social and economic facilities built by local authorities, most of Côte d'Ivoire's elected representatives would be indicted for misappropriation of public funds.

Chapter 7: Local development planning

Local development planning involves drawing up a plan to guide development actions. Development planning means foreseeing, anticipating, organizing and carrying out the actions deemed necessary.

7.1. Objectives and need for planning

In the quest for development, local authorities give themselves a compass or dashboard through the development plan, which clarifies the path to be taken. The plan puts in place the guidelines that circumscribe the framework for intervention. The need for development planning is well established. The aim of the plan is to better target actions, operationalize them and guarantee their effectiveness.

Local development is a collective endeavor requiring the support and contribution of all players operating at local level. For all these players, it means participating in the identification, preparation and implementation of development projects or actions in the various sectors of human activity.

7.2. Avatars of centralized planning in Côte d'Ivoire

Previous centralized development planning was cumbersome and imprecise. The results obtained through this planning always fell short of the expectations of populations, public authorities and development partners. Inequalities between regions were accentuated. Generally speaking, centralized planning had the following shortcomings:

- *Waste of resources because they are misdirected;*
- *Lack of knowledge about the targets of development projects or actions;*
- *Regional planning and equipment based on arbitrary, subjective and questionable criteria;*
- *Lack of realistic strategies for managing development projects or actions ;*
- *Lack of coordination of actions to improve people's well-being.*

7.3. Vision of the community's management team

The notion of local development vision is very important. The vision consists in projecting the community over a given time horizon. A leader might say: "*I'd like my community to be at such and such a level of development in five, ten or twenty years' time*". The local elected official responds to this concern by seeking to provide people with pleasant or decent living conditions and environments. Before this elected official, his predecessors had a vision for the same community. The vision of the new management team is a new one.

The vision of an entire management team is not neutral in itself. It is fundamentally linked to the shared values held dear by the community. These may include the elected official's love of the community, respect for the public good, sharing and solidarity. The vision is legitimate. However, to be valid, the vision must be set against local realities. A vision that is not based on the realities of the community appears to be a mere intellectual exercise, a mere figment of the imagination. In the same way, a vision built outside the community's general environment is a mere figment of the imagination. An elected official's vision for his or her community cannot be thought of in isolation from the people who are the stakeholders in its development.

For the local elected official, one thing is to have a vision for his or her community, even the clearest possible one. Another is to transform this vision into development programs or projects to meet the community's many challenges. The vision must guide all players. To do this, it must be shared with all team members, including collaborators, and above all with the population, in all its different components. Each component of the population must have a sufficiently clear understanding of the vision of the elected representative or leader, and, above all, take ownership of it. A vision that is not appropriated by the population will not bring any results to the community.

7.4. Implementation of the elected representative's global strategy

A community's elected representative or leader sets out his or her vision in a strategic development plan. This involves outlining the way in which he or she intends to carry out, or plans to carry out, the community's development initiatives. This strategy must be presented to the other members of the team for their opinions and contributions. A spirit of collegiality must prevail within the team. The elected representative does not decide alone. He or she decides, in a collegial manner, with the other members of the team and with the community's partners.

7.5. Priority strategic objectives

The priority objectives of the community's strategic development plan are :

- Strengthen the community's national and international positioning;
- Improving community know-how and performance;
- Improve the quality of services offered by the community to the population;
- Organizing people to develop an entrepreneurial culture. This means supporting young people who have legitimate ambitions to become great, by providing them with the tools they need to realize their ambitions and dreams.

7.6. Operational action plan for development

The action plan drawn up by the community must highlight the following elements:

- Ambitious community program;
- Strategic areas of intervention :

- ✓ Reforming the administrative functioning of the community ;
- ✓ Technical reorganization of the community ;
- ✓ Optimizing the community's internal operations ;
- ✓ Remobilization/remotivation of community workers ;
- ✓ Clarifying responsibilities (Who does what?) ;
- ✓ Capacity building for key community personnel ;
- ✓ Strengthening internal communications within the community;
- ✓ Strengthening the mobilization of financial resources ;
- ✓ Controlling the costs of initiatives ;
- ✓ Strengthening the coordination of activities ;
- ✓ Improving community performance;
- ✓ Strengthening community management of partnerships.

Community development planning is not a fad. It's necessary. It's of great importance to the community and its residents, now and in the future.

7.7. Quality of community management

The current management of a large number of local authorities in Côte d'Ivoire remains rather weak, due to a lack of training or capacity building in governance. The quality of local authority management depends, above all, on the sharing of ideas between the manager and his or her staff. Quality management frees up the energy of community workers. It gives them a sense of well-being, and a desire to work well, to excel in their mission.

Through quality management, the local councillor must get the best out of each member of his team. This ability prepares them to give the best of themselves in the conduct of their actions.

Part 3: TOWARDS SUSTAINABLE LOCAL DEVELOPMENT

Chapter 8: City production and governance

In local authorities, the city and its general environment are the result of national development policy. The city remains the receptacle for actions stemming from sectoral development policies. The quality of its production and governance determines the quality of life it offers its inhabitants, as well as its international competitiveness.

8.1. Lack of control over national urban development

8.1.1. Unaccompanied planning

Côte d'Ivoire's cities are, for the most part, spatially in disarray. They are a juxtaposition of pieces called real estate projects, with no coherent link.

8.1.2. Unregulated self-build

Towns in Côte d'Ivoire have always been built mainly by self-builders. In principle, houses are built in compliance with urban planning and architectural standards. But the main rules have not always been respected. This explains the many aberrations we see in the field.

8.1.3. The challenge of urban road management

The urban road network in Côte d'Ivoire has two characteristics: the lack of practicable roads in urban areas, and the deterioration of existing roads due to erosion.

8.2. Failure to manage urban services

The lack of control over solid urban waste management is largely due to the absence of a local and sectoral strategy. This lack of control is characterized by the proliferation of all kinds of waste. Among these wastes, plastics are increasingly visible on urban estates. Plastic waste is an urban resource that can be transformed into paving stones (as the Plateforme des Services in Agboville is already doing) and other products to protect the physical environment of towns and villages.

8.3. Some aspects of urban governance

8.3.1. Structural weaknesses

Analysis of the administrative, institutional, regulatory and decision-making frameworks of urban management has enabled us to draw up the following assessment-diagnosis. With regard to the administrative framework, it can be noted that the decentralization of power management is an irreversible process. It is decentralization from the top down, decentralization without appropriate resources for local authorities. It is also characterized by the weak technical organization of local affairs, by the need for capacity-building for local authority staff, by the absence of training institutions for municipal workers, by the insufficient mobilization of national expertise, and by virtually non-existent popular participation due to the mental unpreparedness of the population to take part in the running of local affairs.

8.3.2. Sharing and managing responsibilities

For a long time, popular participation was ignored in many projects. Recognized as an essential link in urban management, popular participation has yet to be built in Côte d'Ivoire. Decentralization still has a number of shortcomings, particularly as regards the sharing and management of responsibilities and the distribution of resources allocated to local authorities. The issue of transferring powers from the State to local authorities, inter-institutional conflicts, public funding of decentralization and local development and the main tools developed are all aspects that need to be tackled.

8.3.3. Insufficiently operational institutional framework

An assessment of the institutional framework for urban management has been carried out. It shows that urban management is strongly conditioned by the level of functioning of the overall institutional framework. It is based on the diversity of institutional players and the objectives assigned to this framework. However, this framework is marked by instability and inadequate training for the local technicians responsible for running it.

8.3.4. Non-compliance with established rules

Conflicts of jurisdiction that punctuate the life of local institutions and are mostly due to non-compliance with established rules, untimely changes in institutional settings and inadequate communication between institutional players have been identified as factors profoundly affecting urban management.

8.3.5. Lack of organization and coordination strategy

The institutional framework that has been put in place has not yet reached a level of operation that can lay the foundations for sustainable local development. There are still a number of obstacles to overcome, such as poor communication, inadequate technical organization, inadequate training for stakeholders, a lack of initiatives to promote human development, difficulties in mobilizing existing national expertise for local action and, above all, a lack of coordination of development activities. The combination of all these factors prevents the mobilization of the intelligence, strengths and energies required for effective, sustainable local development.

8.3.6. Insufficient enforcement of urban regulations

As regards the regulatory framework for urban management, we note that it is characterized by its inadequacy, that it clashes with popular practices and that it is insufficiently applied. The lack of a strong urban culture is reflected in the inadequate application of urban regulations. The difficulty of integrating the economic activities of the informal economy into the urban space, and a tendency on the part of elected municipal officials to protect their electorate rather than the urban realm, are seen as factors limiting the application of urban regulations.

8.3.7. Weak scientific knowledge of the city

Finally, the decision-making framework for urban management is characterized by poor scientific knowledge of the Ivorian city, and the absence of urban data for decision-making. The decision-making process is globally biased by the absence or inadequacy of reliable information, in a local context where the production of statistics is insignificant or non-existent.

8.4. Abidjan Declaration

At the end of the Abidjan meeting on sustainable cities, held on February 27 and 28, 2020, a declaration, known as the Abidjan Declaration, was made by the participants. On examining this declaration, the following observations can be made:

- It is heavy, vague, opaque and even imprecise;
- It lacked a popular base, because the participants didn't include the residents, the people who make the city, who live the city;
- The participants identified a number of areas for action;
- They forget that cities are not identical and that each axis will have to be adapted to the realities of each city, each country;
- Bordeaux or the France-Africa city summit will do nothing for African cities, because it's not in France that their problems should be posed and solutions envisaged. Moreover, Julien Denormandie warns African governments and local councillors: "The city of tomorrow is being built with the inhabitants, who must have their place in the decision-making process, in consultation and in the implementation of projects" .[5]

African elected officials are responsible for everything: roads, economic infrastructure, various social and community facilities. Elected representatives and local residents must take care of all areas of the city, including public spaces, semi-public spaces, abandoned roads, sidewalks, sidewalks and so on. By raising awareness, elected officials must encourage every inhabitant of the city to take care of a space, in particular the public or semi-public space in front of his or her house, in his or her neighborhood, in his or her district.

8.5. Urban sustainability in Côte d'Ivoire

8.5.1. A paradigm shift

The experience of urban management, which began in the early 80s and continued until the late 90s, has increasingly given way to urban governance. Indeed, in the face of all the ills facing Ivorian cities, the emphasis is now going to be on efficient use of what exists, and on the quality of the inclusive offer of urban services to the population, with a view to seeking efficiency and sustainability on the political, social, cultural, economic, environmental, security and other levels.

8.5.2. A holistic vision

---

[55] Julien Denormandie, French Minister for Urban Affairs, on the occasion of the Abidjan endeavour, from February 27 to 28, statement reported in the daily Fraternité Matin of February 28, 2020 , page 3.

The urban governance approach is based, firstly, on a holistic, systemic and spatially integrated vision of the actions to be taken. Secondly, it seeks to ensure the efficiency of actions by optimizing costs and maximizing results. It also calls for a re-examination or reconfiguration of modes of access to urban services, an imperative for social cohesion, greater citizen participation and strategic integration of actions.

The end result of the urban governance approach is to provide citizens with the means for a good life. The concern of local public authorities is to put in place all the conditions needed to satisfy citizens' essential needs.

8.5.3. A multidisciplinary approach

World Bank experts (1991) assert: "In most countries, the city has become the center of economic and social innovation, culture and political power. The prospects for new urban growth make it all the more urgent to improve the productivity and management of cities"[6] . Good urban governance is a key factor in the social and economic development of nations. It is the key to real and lasting social and economic progress.

Urban governance aims to efficiently mobilize and optimize the intervention of political players, social and economic interest groups, and institutions seeking to carry out urban projects in order to sustainably plan and develop urban territories. Urban governance also calls for the pursuit of results, and the sustainability of those results over time. It lies at the heart of sustainable urban development.

Sustainable urban development could mean the stability of the institutional framework, the reliability of processes, the durability or sustainability of the results of actions carried out in terms of the environment, accessibility to financial resources or the extent of jobs or income-generating economic activities to reduce poverty. It also means the good performance of urban services and facilities, and the good quality of the results obtained, through good urban governance.

With the urban governance approach, city councillors have a leadership role to play in the sustainable development of the whole city, in order to tackle all the problems it faces, without exception. Their work involves coordinating actions and negotiating with the various stakeholders to form a genuine coalition against the problems facing cities. Ultimately, the paradigm of good urban governance is now concerned with rules and mechanisms that are prepared and used effectively, intelligently and coherently by all players, particularly with strong popular participation. In this way, perceived and organized, good urban governance should enable us to achieve convincing and lasting results.

8.6. Optimizing the urban governance approach

[6] **WORLD BANK**. ***Urban policy and economic development: an agenda for the 1990s***. World Bank Policy Brief, 1991, Washington, p. 23.

8.6.1. Clearly defined objectives

The urban governance approach defines the prerequisites for successful urban development and a pleasant living environment. It enables us to define a modern, integrated urban policy that offers citizens a wide range of opportunities. It is based on clearly and consensually defined development objectives, and a transparent approach to doing business based on a clear division of responsibilities. A city that is well governed politically, administratively, technically, environmentally and financially, has a positive and definite influence on the living conditions of its inhabitants.

Optimizing urban governance can only be achieved if certain conditions are established in advance. These conditions include political, social, institutional, cultural, economic, environmental and security aspects. It is around these different aspects that it is built.

8.6.2. Strong political commitment

To produce the expected results, urban governance requires a strong commitment from all players. The absence of such commitment sets the stage for failure and development. The commitment of each player or group of players is essential. This means full commitment, from start to finish, to all the processes involved. It's not a question of forcing each actor or group of actors to assume responsibilities, but of getting them to become aware of them themselves and to commit themselves freely, firmly and sustainably.

8.6.3. Mobilizing stakeholders

The commitment or willingness of community players to work towards good urban governance is an important step. But it's not enough, because it doesn't guarantee anything. It must be accompanied by the unfailing mobilization of all players. Mobilization is far from simple. It is even difficult to achieve, because of prejudices, biases and contexts, some of which are not necessarily favorable. Appropriate strategies can therefore be developed to ensure this mobilization, which is at once ideological, political, social and economic.

8.6.4. Quality of urban knowledge

The urban governance approach is based on the principle that the formulation of urban development policies, programs and projects depends intrinsically on the existence of reliable, up-to-date data. The improvement of urban knowledge can be achieved by anticipation through research or prospective studies.

8.6.5. Quality of urban services

Urban populations are increasingly demanding quality public services. The urban governance approach, which is concerned with the smooth running of services such as security, transport or markets, places particular emphasis on the measures to be adopted to enable municipal services or concessionaires to play their role properly. Henceforth, the quest for quality in the urban services offered to the population is a priority objective for suppliers. In order to achieve this sustainability, central and local public authorities must each work towards the "*programmed and integrated implementation of infrastructure, its operation and maintenance*" [UN-HABITAT, 2012] . [7]

In terms of urban safety, for example, political, administrative and municipal leaders must work to strengthen safety indicators. These indices can be summed up as the possibility for residents to stroll freely in their neighborhoods and towns, to rest in a public or wooded area, to walk and circulate freely on sidewalks, without the risk of being run over by reckless motorists.

It also means that residents can alert the police in the event of aggression, and receive a response within a reasonable timeframe. All in all, if urban sustainability is to be effectively achieved, everything must be done to ensure that this quality is attained. With this in mind, partnerships can be developed with the public and private sectors to improve the quality of service provision.

8.6.6. Asset maintenance

The upkeep and maintenance of urban assets has not always figured prominently in urban policies in Côte d'Ivoire. Now, with the urban governance approach, the protection and maintenance of urban heritage occupies a major place, notably through the programming, realization, operation and maintenance of infrastructures (infrastructure and superstructure equipment) by public authorities, all based on an efficient and sustainable financing mechanism within the framework, for example, of public-private partnerships (PPP). At the same time, we need to promote a new type of citizen, one that is disciplined and aware of the need for good management of our shared urban heritage.

8.6.7 Responsible territorial governance

Responsible urban territorial governance is advocated by all. It calls for a "responsible" awareness on the part of all actors and users of the urban domain to maintain it in a satisfactory state of sanitation. In Côte d'Ivoire, this can be achieved through :

- i) the promotion of a mode of production and spatial development of the city, compatible with the long term (future generation) and placing the inhabitant at the heart of the action;
- ii) coherence between the various physical components to ensure harmonious and optimal functioning of the urban domain, in particular by enabling each spatial entity of the city to perform its vital functions and ensuring fluid mobility for all populations, from one component to another;
- iii) control over spatial governance by combating the disorderly expansion and occupation of urban areas by local populations, provided that all of them understand the importance and scope of a healthy urban environment.

8.6.8. Preparing the shared urban future

[7] State of the world's cities 2012/2013- Prospirity of Cities, UN-HABITAT, Nairobi, 2012, 151 pages.

The urban future of Côte d'Ivoire is realized or woven from the present urban. The urban future, or the future state of the urban community, is not a foregone conclusion. It will be the result or combination of all the actions that will be planned, programmed and carried out, from now until the chosen time horizon.

Preparing for the future requires foresight. It also requires the adoption of strategic mechanisms or approaches in every area of urban activity, with, of course, the use or application of information technology and e-governance to all aspects of municipal management for diligent, efficient and cost-effective service delivery.

Today, it's no exaggeration to say that most Ivorian cities are deficient in terms of their planning, production, organization and governance policies. Overall reflection on Ivorian cities is generally lacking. In this context, its production does not follow a coherent framework. Public authorities have yet to mobilize the human and financial resources needed to govern Ivorian cities. To this relatively gloomy picture, we must add the approximate management of the urban domain. All in all, the Ivorian city still faces many challenges.

Chapter 9: Change for local development

Côte d'Ivoire is changing. Change is seen through the mutations taking place on the political, social, demographic, cultural, economic, environmental, climatic and other levels. Faced with these changes, we need to prepare political players and local authority managers to propose new visions, develop new capacities, new approaches to intervention, new governance, new strategies, and propose new behaviors.

9.1. Names of local authorities

The issue of local development in Côte d'Ivoire has always been characterized by changes in the names or boundaries of local authorities for political or party political purposes. This has led to a degree of administrative instability in the running of these communities. This instability has prevented development from taking place on a realistic administrative basis.

We don't want to dwell on the underlying reasons for these name changes, or on the consequences of these changes or divisions on the operation of local authorities. In all cases, the untimely division of local authority territories requires readjustment to local realities, particularly in political, human and economic terms. For example, there is the question of the physical accessibility of certain local authorities as a result of administrative redistribution without objective justification. There are also frustrations, even discontent, felt but contained by a large proportion of the population, as a result of being forced to belong to a new administrative entity. People find it difficult to identify with this new administrative entity.

Faced with this situation, the local population is unanimous in its demand for greater stability in the local administrative landscape. For local populations, the administrative concerns resulting from a whimsical division of their territory are, in reality, a psychological brake on their participation in local development initiatives. The stability of local authority territories could facilitate the implementation of major infrastructure and equipment projects, and thus enable them to kick-start development.

9.2. Enhancing the current role of local councillors

The office of mayor, for example, has changed little in Côte d'Ivoire from 1980 to the present day. It has remained virtually static over the period. All functions evolve over time, as local human society evolves. Over the years, local authorities have evolved, or at least undergone more or less profound changes, notably in political, demographic, spatial, socio-cultural, environmental and economic terms.

The role of the local elected official is to lead the community's development programs and projects. There is a clear gap between the role of mayor and the progress of the municipal institution, particularly in terms of taking on new responsibilities. Côte d'Ivoire's mayors are calling on the public authorities to increase their allowances, currently considered derisory, to one million CFA francs. For large and medium-sized communes, it is possible to grant mayors this allowance. But for small communes with modest budgets, such an allowance is unlikely to be granted. Unless the public authorities decide to grant a living wage to the mayors of small communes.

9.3. Conflicts of competence between directors and elected representatives

In Côte d'Ivoire, the introduction of decentralization for local development still leaves plenty of room for devolved administration and decentralized state services. Their role is to act as an interface between the central administration or authority and local authorities. In carrying out this mission, central administration and decentralized services must provide local authority managers with support and, above all, strategic guidance to ensure better understanding and, above all, optimal use of planning, administrative, technical and financial management tools.

For newly elected local government officials, mastery of the procedures, methods and techniques of planning, programming and budgeting, in line with legal and regulatory provisions, is an essential prerequisite for carrying out their mandate. Without such mastery, it is virtually impossible for elected representatives to effectively manage development projects or actions in their communities. Today, there are still conflicts of competence between the different entities, due to the non-clarification of aspects linked to the transfer of competences.

In Côte d'Ivoire, the management of development initiatives by local authorities is hampered by the persistence of these jurisdictional conflicts. These conflicts concern, among other things, the construction of allotments, particularly in villages that are not located within a communal territory. Even in communal villages, the conflicts are obvious, as subdivisions generate lots and therefore financial resources. Conflicts of jurisdiction also arise when entities are represented at official ceremonies. Faced with these situations, the public authorities need to clarify the roles and responsibilities of all local players.

9.4. Clarifying the status of local authority employees

For the conduct of activities by local authorities, a fairly rigid organic framework of jobs has been adopted by the supervisory authority. Local councillors are required to comply. This rigidity leaves them little room for manoeuvre when it comes to recruitment.

The mobility of local authority staff remains a problem that needs to be clarified. In town halls and regional councils, the staff most affected by mobility are heads of technical departments. Some of them leave because of the incompatibilities that arise between the elected official and the technician. This creates an unhealthy climate that prevents frank and close collaboration. As a result, the technician is forced to ask for an assignment to pursue his or her career in another local authority. Local councillors usually justify their decision to let a technician go on the grounds of insubordination.

9.5. Lack of a framework for reflection

In Côte d'Ivoire, local authorities do not have, or are not equipped with, a strategic thinking framework for their smooth running. In most cases, elected representatives and their staff are content to manage day-to-day operations or routine. That's all they do. "*Look at the piles of documents yourself. I'm the only one who analyzes them*"[8] , Ruffin Pouho, then head of technical services at Daloa town hall, confided to us in 2017. The sight of documents piled one on top of the other, on tables, is enough to convince us of the impossibility of organizing strategic thinking within the technical team.

It should also be noted that within municipal technical departments, there is a senior manager (generally a public works design engineer for large town halls, especially regional capitals, a public works technical engineer for departmental capitals, or a senior technician for small communes). In almost all Ivorian local authorities, there is no other technician with the same level of education. This is one of the major handicaps of Côte d'Ivoire's municipal technical services. This situation makes it impossible to set up a framework for optimizing the operation of these services.

Faced with the weakness of municipal services in fulfilling their responsibilities, community initiatives are springing up. They try to compensate for this weakness by providing various services. These services are provided directly to local residents on behalf of the municipal administration. In principle, the service providers are remunerated by the municipal administration, which is legally responsible for providing these various services to the local population. In some cases, service providers are paid by local residents, who are the direct beneficiaries of the services.

9.6. Insufficient control over land management

Mastering land management, in both urban and rural areas, is of strategic importance to communities. Land is their main means of survival. So it's easy to understand why communities are so determined to defend their land. Land is their life, and it remains an essential concern. Land management in Côte d'Ivoire is characterized by the following aspects:

- Land disputes are on the increase, as several social and economic players compete for land in local communities. In this battle for control over land, not all players use the same legal means;
- Slow delivery of administrative acts by the relevant departments (lack of promptness);
- Slow handling of disputes by the administration due to the diversity and complexity of situations;
- Poor internal circulation of information. There are cases of deliberate withholding of information by certain agents of the administration for personal and not necessarily admitted motives. The withholding of information constitutes an abuse of power on the part of the agent who does it;

[8] See date in my notes on Daloa.

- Conflicts of competence between professionals in the land sector, some of whom behave in ways that don't comply with current regulations, generally for corporate interests.

9.7 Inefficient management of the urban estate

Urban and village estates all over the country are disjointed and untidy. This disorder is mind-boggling. It has been building up for decades, in full view of the public authorities and established institutions, sometimes with their agreement or complicity. Everyone is aware of the impact of an illegal administrative act. But no one has really worried about it.

In this context, self-building, the leading method of producing urban space in Côte d'Ivoire, had not been assisted by the public authorities and their branches. The lack of respect for even the most basic building regulations was blatant. This approach had been encouraged by the laxity and laissez-faire attitude of central and local public players. The effects of this unassisted self-build were disastrous for the public domain.

The public, semi-public or decentralized administration has not been able to deploy itself, in a suitable manner, on the national territory in order to lead or assist the construction of towns and villages. This situation largely explains the poor image of urban areas in Côte d'Ivoire.

9.8. Community performance assessment

Assessing the performance of local authorities is a necessity. It represents a strategic challenge. It creates the conditions for emulation and progress among the country's local authorities. In particular, it lets them know that they are in competition with each other, and that they need to give their best. Finally, international local development organizations know which communities they can work with.

Assessing the performance of local authorities is an activity that requires political will. It is the political will that sets the rules of the game, and in particular the objectives of the evaluation. At present, this political will does not exist.

The choice of evaluator is also very important. He or she must be an independent player. The scope of the evaluation, the evaluation methods, the evaluation indicators, the practical organization of the evaluation and the final use of the evaluation products are all aspects that need to be made explicit.

9.9 Socio-professional integration of young people

In Ivorian local authorities, young people (aged between 15 and 35) account for 60% of the population (RGPH 2014). Youth employment is a preoccupying issue. It is certainly one of the very first challenges facing local authorities. Today, employability is at the heart of the local debate. The concept of employability can be understood as the possibility for a young graduate to access a job in line with his or her qualifications. This raises the question of the training-employment match. It also raises questions about the curricula used in training institutions.

In addition, access to employment extends to self-employment, i.e. the ability of graduates to create their own businesses. Self-employment is not the sole responsibility of schools and training institutions. The family or family unit, for example, must play an active role. Likewise, religious and non-governmental organizations have their share of responsibility in creating a breed of young entrepreneurs who are ambitious, serious, conscientious and hard-working, and who don't confuse capital with profit.

Identifying employment potential or opportunities for social and economic integration, particularly for young people and women, is an activity that needs to be taken into account. It must be given the utmost attention. The question of employment or self-employment for young people needs to be examined throughout their school or university career, right up to the moment they graduate. This means intelligently linking teaching and apprenticeship, particularly in work-study schemes. Teaching and on-the-job training must be carried out simultaneously. This examination will determine whether the graduate is capable of self-employment.

Training young people in the creation and management of income-generating activities (administrative management, technical management, financial management) is an equally important activity. Within local authorities, this issue must be given pride of place.

Chapter 10: Universities for local development

During the local election campaign, I listened to the radio, watched television and read what many candidates had to say. They all talked about the development of their communes and regions. What astonished me most was that none of them, in the regions that have them, mentioned what the local university can contribute to the progress of the commune or region. I found it abnormal that candidates in local elections should ignore universities and the role they can play in the formulation and implementation of regional or local development policies, programs or projects.

10.1. University: a key player in local development

Reflecting on this observation, I asked myself the following question: "*Is it really the fault of the candidates in the local election*? I made a hypothesis and said to myself: "*It's certainly the fault of the university or universities that have not been able to impose themselves locally, apart from teaching the students, the children of the regions and of Côte d'Ivoire*".

The fact is that universities have not been able to impose themselves on national and local decision-makers as essential players when it comes to making decisions on local development policies, programs and projects. "*What should be done about it*? The only thing to remember is that the university, wherever it is located, must be seen as a key player in local development.

10.2. The role of the university

The main mission of the university, or any university, is to train managers specializing in local, rural and community development. More specifically, it must :

✓ ***Teaching, awarding diplomas, degrees and titles***

One of the main roles of the university, or any university, is to provide teaching in different disciplines, award diplomas to learners in the various cycles, and grant degrees and titles to teachers and researchers.

✓ ***Producing scientific knowledge.***

Academics talk about scientific production. Scientific production is a process. Scientific research, or scientific research activity, starts at point A and ends at point F. Between the two points, there are a multitude of questions or issues that the researcher or team of researchers must answer. When it comes to scientific research, there are, of course, research policies, programs and projects. There are also research teams to set up, technical resources to mobilize for scientific research (notably equipped laboratories), financial resources and research methodologies. Research projects and activities aim to enlighten decision-makers and development partners on the problems facing a nation or region.

A good understanding of problems through research provides decision-makers with the objective information they need to make the right decisions. What guides the researcher or team of researchers is, above all, the quest for quality in scientific information.

Scientific research leads to results or products. It is the results or products of research that feed into the content of the teaching provided.

✓ ***Disseminating scientific knowledge***

For the university, it's not about producing knowledge for the sake of producing knowledge. It's about producing scientific knowledge or information to answer the questions facing a region or a nation. Scientific information or knowledge, insofar as it facilitates decision-making, must be accessible to all those who request it. There is talk of democratizing the use of scientific information. Where does or can scientific information be disseminated? Generally speaking, scientific information is disseminated at two main levels:

➢ **Within the** university **institution**

This dissemination can take the form of the exchange of scientific reports or documents between departments, research centers or institutes, or laboratories. It can also be achieved through the presentation of the fruits or products of research work at seminars organized on site.

In 1994, I went to study the environment at a small university in England. It had just over 8,000 students. Today, it has over 20,000, according to information gathered from the university's website. The town of Louborough has a population of around 60,000. It's a small industrial town some thirty kilometers from Leicester, the capital of the Leicestershire region. It used to be called the University of Technology. What I found at Lougborough University was that practically every day there was a conference in some department or institute. Teachers, researchers and students were invited by posters or on the local radio (the university has its own local radio station) to take part in these conferences. There was a real, genuine and constant scientific liveliness in the university. I was deeply impressed by the lack of scientific activity at Félix Houphouët-Boigny University. There are no research activities, due to a lack of research funding.

Scientific activity within a university is more intense where scientific research is funded and research teams are constantly on the go. Under these conditions, there's scientific information to be disseminated and shared, all the time.

➢ **Outside the University.**

- ✓ Scientific information can also be disseminated between the region's university and other universities in Côte d'Ivoire and abroad. A university cannot live in isolation. It needs to be in touch with other universities, both inside and outside the region.

- ✓ The dissemination of scientific information can also take place between the university and local industry or business. The reason why this type of exchange does not take place today is that industrialists are not involved in defining research programs or funding research, for example. In fact, they are not invited by universities. In future, if local industry meets this condition, it will be closely involved in defining research and training programs. A closer look at the situation reveals a lack of understanding between academia and industry. The reality is that the region's industrial fabric itself is insufficiently developed. In fact, it's still in its infancy. As a result, it is not yet in demand for scientific information.

- ✓ Scientific information can be disseminated between the university and the local community as a whole. The university must be open to the whole of local society. Every citizen, every organization in society, every public or private institution can turn to the university to satisfy a specific need for knowledge, with a view to facilitating decision-making. This is the democratization of the use of scientific information. Disseminating or transmitting information to the public helps to raise awareness of the fact that the university is at their disposal.

10.3. Some local development requirements

Local development has a number of requirements, three (3) of which are outlined below:

✓ ***Rigorous basic organization***

For local development to be real, effective and sustainable, activities must be rigorously organized. This organization, in turn, must be based on a method or approach with clear objectives and an effective strategy for running development programs or projects.

✓ ***Enlightened leadership***

Local development unquestionably calls for leadership. Before being a matter for the whole community, local development must be imagined and thought out by one or more leaders. Where there is good leadership, local development is well conducted. The absence of enlightened leadership can prevent a region's economy from taking off, even though resources are available and even abundant.

✓ ***Better understanding of local realities***

Managing the development of a locality or region requires a thorough understanding of local or regional realities. These realities need to be known in minute detail.

10.4. Sustainable local development concerns

Sustainable local development is development that provides solutions for a better life not only for current generations, but also for future generations. As such, it is based on socio-cultural, economic and environmental or ecological dimensions.

- ✓ ***Social and cultural concerns***

  Taking into account the region's socio-cultural concerns is essential. Sustainable local development puts people at the heart of its actions.

- ✓ ***Economic concerns***

  Economic concerns must also be taken into account. Sustainable local development is based on long-term economic activities.

- ✓ ***Environmental concerns***

  Taking environmental or ecological concerns into account is essential. Sustainable local development places the environment or ecology at the heart of its actions.

10.5. University contributions to local development

- ✓ **Initial and continuing training for managers and technicians**

- ➢ It provides initial training for the managers and technicians needed by the local economy (industry, energy, agriculture, commerce, services). The question of matching training to jobs (or employability) is omnipresent. The responsibility of both the public and private sectors is engaged. The private sector must participate in defining curricula, indicating the profiles that are or will be required by local industry.

- ➢ It provides ongoing training for managers and technicians, giving workers in various sectors the opportunity to return to university to enhance their operational capabilities by acquiring new knowledge. These programs must be adapted not only to the needs of public sector workers, but also to those in the private and even voluntary sectors. These programs should be implemented in a flexible way, in the form of evening classes.

- ✓ **Knowledge of local realities**

Various studies are required to highlight the essential characteristics of the region. This involves taking a snapshot of the region's natural, political, social, cultural, economic and environmental realities. It should be noted by local elected officials that the formulation of sectoral policies, development programs and projects must be based on intrinsic knowledge of the region's realities.

10.6 The university's contribution to local development

✓ ***Produce strategic and forward-looking thinking***

Any local development policy, program or project must be supported by strategic and forward-looking thinking, which provides local decision-makers with objective information to help them make decisions. This strategic and forward-looking analysis must be permanent and holistic, i.e. it must cover all the region's problems. It can be carried out by local university experts, or by independent consultants.

✓ ***Working in synergy with local industry***

The local university must be at the service of the local industrial community. Through its presence, and depending on its vocation, it can help attract industrialists to the region. To achieve this, the university must adapt its training courses to the needs of local industry. There must be a real synergy between local industrial activity on the one hand, and scientific activity, teaching and training on the other. Local industry and the local university need to work together, in other words, in synergy. They need to learn to work and win together. It's a combination that can make the whole region win. We absolutely have to try it.

✓ ***Working frankly and confidently***

Local government leaders must work with scientists and experts from local universities in a spirit of openness and trust. All actions to be initiated by both parties must be based on this openness and trust. The solidarity and sustainability of cooperation with local authorities, for example, must be based on truth and trust.

✓ ***Working with multidisciplinary teams***

Development is a whole. It has many dimensions. To cover all these dimensions, the local university must offer multidisciplinary, committed and competitive teams.

✓ ***Working together in humility***

Local authorities and the local university need to work together to meet the many challenges facing the region. This cooperation can only be achieved with humility. Humility guarantees the success of any initiative to cooperate with scientists and experts from the local university.

✓ ***Work on the preparation of project documents***

Preparing project documents is an essential task for local authorities when managing their cooperation or partnership. Local authorities' project documents must be technically well prepared, well-crafted and bankable. The credibility of local authorities depends on the quality of the project documents they present to actual or potential partners.

It's generally said that money is everywhere. What's missing or lacking, more often than not, are well-crafted project documents that meet the quality requirements of development funding organizations or agencies.

Furthermore, local councillors are asked never to go and meet an organization without a project document in hand. In 1989, during a training course in Brussels (Belgium), a United Nations (Habitat Nairobi) expert by the name of Nicolas You came to give us a lecture. He told us: "*Most Africans go to forums, colloquia and the like, empty-handed, whereas others go with well-prepared project documents*". In 2020, a local elected official cannot go and meet the heads of national, sub-regional or international organizations with a simple list of projects, as some tend to do. You can't do anything with a list of projects.

The presence of a university or universities in a region is an opportunity that should be seized to lead strategic thinking, organization and activities aimed at laying the foundations for sustainable local development. The university must be at the service of the region's communities.

Chapter 11: Training and capacity building

In all sectors of economic activity, training men and women is the best way to boost progress. This is particularly true for local authorities. It is therefore clear that sustained attention must be paid to training and capacity-building for local government players.

11.1. No training is too much

As part of the activities of the Université des Collectivités d'Abidjan, training sessions, generally organized over three days, were offered on various themes linked to local development. Municipal councillors were formally invited to attend these training sessions. However, the latter hardly ever attended these training sessions. For the most part, they preferred to be represented at these sessions by collaborators. This was a serious mistake on the part of elected representatives, who did not take advantage of the opportunities offered to them to strengthen their capacity to manage and coordinate activities within their communities. Training sessions were always led by high-level national experts on the various aspects of local governance. However, due to funding problems, training and capacity-building sessions are not offered all the time.

The mayors' attitude can be explained by a lack of humility. Most of them believe, quite wrongly, that they know everything there is to know about running a municipal administration. A training or capacity-building session is never too much. It always enables participants to increase or improve their ability to understand, lead and coordinate activities during their term of office.

11.2. The need for training and capacity building

There are a number of reasons why it is necessary to train and build the capacity of elected representatives in Ivorian local authorities. We propose three of them:

- Local elected representatives come from a variety of disciplinary backgrounds. The disciplinary horizon of many local councillors in Côte d'Ivoire is not decentralization or local governance;

- A very small minority of local councillors in Côte d'Ivoire have backgrounds in civil engineering, architecture, urban planning, development, geography, project management or economics. On the whole, they have a better chance of succeeding in running their community's activities. The situation is different for the majority of local councillors, who lack the minimum knowledge required. This points to the need for training and capacity-building in decentralization and local development.

- For all local councillors in Côte d'Ivoire, it is necessary to enable them to acquire the knowledge, skills, know-how and interpersonal skills they need to successfully carry out their mission.

11.3 Training for elected representatives in various fields

11.3.1. A strategic challenge

Training and capacity building for local councillors enables them to improve their individual performance and provide effective leadership in the management of their mandate. This represents a strategic challenge for the proper management of the community. Their sound knowledge of mechanisms, procedure manuals and implementation strategies provides a solid foundation for the smooth running of the community. The idea is not always to say, this is the staff's job. When the elected official has a good understanding of the various aspects of his or her mission, he or she is more inclined to give appropriate instructions or guidance to his or her staff. The results obtained are generally better.

11.3.2. Training on several themes

Training and capacity-building for local councillors generally cover a wide range of topics, given the diversity of skills and responsibilities transferred to local authorities. These include

- The aim of training elected representatives is to give them a better understanding of the texts and the powers and responsibilities transferred to them. A good knowledge of the texts governing the organization and operation of local authorities is a very important step. With this knowledge, local elected officials know what to do and what not to do in carrying out their mandate. In short, a sound knowledge of the texts and procedures will enable them to avoid the mistakes that are generally made by many local councillors;

- Training elected representatives in local development initiatives, particularly project management. Local authorities initiate development projects in their various areas of competence. It is the project owner, i.e., the entity that carries out the work. This status gives it responsibilities to assume in the execution of these projects. The project owner must be familiar with the various stages in the project realization process (stages in the project life cycle). And, above all, at each stage of the project life cycle, he or she must know what documents need to be produced and given to the various bodies involved in the project;

- Training local elected representatives in environmental, sustainable development and climate change issues is essential. Today, sustainable development and climate change are central to local development policies, programs, projects and actions. Local councillors must not shy away from these new paradigms. Better still, they must be prepared to play their full part in controlling climate change;

- Training elected municipal officials to optimize urban solid waste management, to improve the quality of the urban environment. Training or capacity building of local elected officials will enable them to better respond to the challenges of solid waste;

- Training elected representatives in local authority financial management is of the utmost importance. Elected representatives' mastery of financial management procedures and mechanisms is an asset;

- Training and capacity-building for local councillors in optimizing tax collection will enable them to better understand the mechanism and increase their local authority's financial resources;

- Training elected officials in contract management (construction work, supply of goods and services and other services), enables them to master the preparation of technical files and tender documents, as part of the project management of development projects;

- Training elected officials in the development and management of partnerships is of strategic importance to the community. It enables elected officials to make the most of partnerships forged with institutional players from outside the community;

- Training and capacity-building for elected municipal officials in the management of urban public property is one of the major challenges of territorial governance;

- Training elected representatives in the introduction of digital technology into the management of community activities and services in the future will enable them to stay one step ahead of the upheavals brought about by digital technology.

There is no limit to the number of training and capacity-building initiatives for local elected representatives. Today, the question is how to finance training and capacity-building initiatives. To date, public authorities have not yet understood the importance and necessity of investing in training and capacity-building for local councillors and their staff. The lack of training is an obstacle to preparing the future of local authorities and their populations.

11.4. Proposed teaching methods

During training and capacity-building sessions, local councillors have the opportunity to share experiences on practical cases. In carrying out their mandates, local councillors work with colleagues on a wide range of issues. They do not always have the opportunity to share their colleagues' experiences and draw lessons for their own work.

Sharing experiences with each other is an excellent opportunity for mutual enrichment and preparation for a fresh start in running their community.

Cases of success and failure are presented, analyzed and discussed. At the end of the discussions, a course of action is identified and "validated". Elected representatives, thus trained, return home, as it were, armed, reinvigorated or rejuvenated and ready to face the challenges ahead, with greater enthusiasm and self-confidence. The training or capacity-building programs will then have achieved their objectives.

Chapter 12: Human resources

The development and performance of local authorities is fundamentally dependent on the availability and mobilization of a sufficient number of high-quality human resources. Their survival and competitiveness depend on the quality of their human resources.

12.1. Quality human resources: a major asset

Education and training, professional experience and know-how are factors that determine the quality of human resources. Human resources development is an essential pillar of local authority management. The availability of quality human resources at local authority level is a factor in development and competitiveness. It is therefore up to local authority managers to put in place an appropriate strategy for developing and mobilizing adequate human resources.

For the time being, most local authorities in Côte d'Ivoire suffer from insufficient and low-quality human resources. This means that they are unable to meet all their organizational and management needs.

12.2. Recruit, employ and retain the best

The elected representatives of local authorities work in a cycle when it comes to human resources management. The cycle begins with the expression of a specific human resources need. To specify this need, the profile of the worker required is defined. The worker or agent recruited is employed within the local authority. He or she is assigned a function. He or she must demonstrate certain values, notably a love of work, diligence, respect for the public good and the pursuit of individual excellence. This is where the best people come into their own. The concern of the local councillor is to retain the best people to achieve the expected results.

12.3. Employee recognition

The texts governing the operation of local authorities in Côte d'Ivoire make no provision for specific incentives for staff. The salary scale in force in Ivorian local authorities, it cannot be repeated often enough, does not offer competitive remuneration. Nor do workers receive a performance bonus to encourage them to work harder. What's more, local authority workers' wages are paid irregularly. This does not put them in a good position to fulfill their responsibilities.

12.4 Create the conditions for professional mobility

The professional mobility of local authority workers enables them to change physical, social and cultural environments during their careers. This mobility is necessary. It benefits workers first and foremost, as well as local authorities.

Local councillors must offer their employees the opportunity to move from one local authority to another, without difficulty. For local authority workers, this is not just a change of scenery, but a change provided for by the legislator, enabling them to re-mobilize their professional potential.

12.5. Accountability: a vital necessity

To date, the elected representatives of local authorities in Côte d'Ivoire have no obligation to achieve results. There is no clearly established official requirement in this respect. The absence of an obligation to produce results is confining many local authorities to a certain degree of immobility, monotony and even lethargy. In reality, many local councillors are content with routine actions. This trend needs to be reversed, with elected representatives held to account by central government officials, voters and non-voters, and local development partners.

The obligation to deliver results requires elected representatives to mobilize all the community's resources in the search for appropriate solutions to the problems facing the population. It leads them to harness all their intelligence and energies to drive development initiatives.

12.6. Municipal or regional councillors

Ivorian local authority councillors deserve our special attention. The question the average citizen generally asks is: who are the councillors and what do they do? Citizens need to know exactly what councillors do, how they do their work and what impact their work has on the community. In reality, the advisor's mission is poorly understood by the general public. Above all, citizens need to know the basis or criteria on which municipal or regional councillors are chosen.

The absence of any truly adversarial debate at certain municipal council sessions, for example, suggests that councillors are content to pay their attendance fees. Local councillors are barely visible in their neighborhoods or communes. What local resident has ever seen a local councillor in his or her ward (even if he or she lives there?). It's a sad reality. And it raises the question of the legitimacy of their mission.

Faced with the citizen's many questions, changes are needed. One of the changes to be made would be to overhaul this function to make it more effective, more attractive and more beneficial to the community. We also need to review the way in which councillors are selected or appointed. In addition, for the selection of municipal or regional councillors, latitude could be given to grassroots communities to participate and give them the necessary legitimacy.

The problem is simple. Within communities, it's pretty clear who's who. In other words, we know who has a particular interest in the life of the community. Who is in a position to mobilize the other members of the community to carry out the actions set out in the program?

Finally, to ensure the effectiveness of their actions, councillors should be granted a basic monthly or annual allowance. An annual bonus could be awarded to the advisor who has been active and effectively mobilized community members in community-wide actions.

12.7. Train and support local players

The social and economic players who wish to become involved in local development initiatives need to be identified, trained and supported. In Côte d'Ivoire, the main players in economic production are still those in rural areas. Farmers do not yet benefit from the training they need to boost their production potential. Training small farmers is a vital necessity. It consists in helping them acquire the knowledge and know-how they need to improve their performance and productivity, by producing more, better and better.

Untrained, small farmers continue to farm with the same philosophy, using old, archaic tools that don't allow them to improve their productivity and yield. There is also the question of their overall organization. Poorly organized or unorganized, unassisted and with no mastery of the marketing chains for their products, small-scale farmers are unable to make the most of their work. Speculators are there to make merry on their backs, buying their products at low prices and then selling them at higher prices in urban centers or abroad. Beyond the political and demagogic rhetoric we hear, there's no one to come to their aid, more concretely.

12.8. Expand the role of community leaders

In both rural and urban areas, the role of community leaders remains essential. This technical role needs to be expanded. This would make it possible to:

- ensure the necessary interface between, for example, municipal councillors or regional councillors and resident urban or rural communities or socio-economic players living and working in the community;
- coordinating or supervising collective actions undertaken by resident communities or socio-economic players, e.g. in the areas of public health, noise abatement, combating violence or insecurity within the community, social cohesion and dialogue, etc.

12.9. Civic involvement at work

The quest for local development must be based on citizens' commitment to hard, persevering work. Today, however, it is clear that Ivorians generally do not work hard enough. A spring has broken in the work culture of traditional Ivorian communities. Today, many Ivorians, especially those living in urban areas, feel that work is no longer "enough" to feed their families, that work no longer "pays", and so on. In their view, something other than work is needed, in particular relying on "relationships or knowledge" to exist and evolve in society. It's a negative and dangerous way of thinking, generalized to all communities and the whole of Côte d'Ivoire. In a way, it drags the local community down. And social and economic progress is hard to achieve.

Chapter 13: Financing local development

The issue of financing local development is central. It is of strategic importance for local authorities. It covers a number of aspects, including procedures, local finance mechanisms and budget execution control. These elements are set out in documents drawn up by the supervisory authorities to ensure the sound financial management of local authorities. A good understanding of these elements enables elected representatives to avoid certain errors. Our analysis of the question of financing will focus on four (4) aspects: i) financing from public resources; ii) financing from the mobilization of own resources; iii) financing from specific funds; and iv) financing through the development of partnerships.

13.1. Financing from public resources

The financial challenges of decentralization in Côte d'Ivoire are enormous, given the scale and diversity of the powers transferred to local authorities. These local entities do not have the financial resources required to carry out development projects. All local elected representatives in Côte d'Ivoire would agree.

Financial resources from the State remain largely dependent on its goodwill. In other words, the State alone decides to give local authorities what it wants and when it wants it. This is not negotiable.

For many of us, the lack of funding for local development appears to be an obstacle to our nation's progress. The question of financing local development raises a paradox. On the one hand, there are abundant natural resources in local communities, including land, water and raw materials, in almost every region of the country. On the other hand, there is a deplorable absence or shortage of financial resources to meet the many needs expressed by communities and local authorities. This paradoxical situation is hampering the realization of community development programs and projects.

A recurring issue for local authorities is the need for a single fund. In simple terms, a single treasury means the existence of a common treasury into which central and decentralized public services pay their pooled financial resources. The common fund is the public treasury. The fact is, once the financial resources have been paid into the treasury, it is virtually impossible for local authorities to know, in real time, what share of the resources is theirs. This makes it difficult for them to carry out local development projects.

All in all, the subsidies granted by the Ivorian state to local authorities are notoriously inadequate. At the opening of the awareness-raising and information seminar for representatives of local authorities, on June 27, 2019 in Abidjan, the representative of the Minister of Economy and Finance, Akpess Yapo Bernard, revealed that in 2018, the State granted an amount of 160.2 billion FCFA, made up of 97.9 billion FCFA for retroceded tax quotas and **62.3 billion FCFA for subsidies granted for the year 2018**". These subsidies represent 0.9% of the national budget, estimated at 7653.3 billion CFA francs. In principle, they should represent 10% of the national budget. This was the major recommendation of the review seminar on decentralization held in 2013. During the seminar, participants made the following observations and recommendations:

- the gradual extension of local authorities' tax powers, in compliance with legally prescribed provisions;
- irregular cash flow to local authorities;
- village or district chiefs and the local population do not take ownership of the texts governing local authorities;
- local authorities not taking responsibility for solid waste management;
- the definition of a national decentralization policy that would be explained so that all components of Ivorian society have the best possible understanding of it, and above all that they take ownership of it and participate in its implementation.

To these various aspects, we must add the setting up, in certain cases, of a parallel organization for the collection and payment of financial resources collected from merchants and economic operators. With such an organization, the financial resources mobilized naturally go into the coffers of an individual or group of individuals. This illegal practice, observed in communities in crisis, deals a heavy blow to the financing of local development.

In the light of this assessment, we note that there are still many challenges to be met in order to effectively achieve a more responsible decentralization that will bring greater well-being to local populations. The debate on the financing of Ivorian local authorities from public resources is far from over.

13.2 Financing from own resources

All local teams are involved in financing local authorities by mobilizing their own resources. The organic framework for mobilizing financial resources gives an indication of the possibilities for developing own resources. The mobilization of own resources depends on the economic, and therefore financial and fiscal, base of the local authority. The fact is that most local authorities are content with what the organic framework suggests. There are other financing possibilities that have yet to be exploited. Efforts therefore need to be made to broaden the tax base and strengthen the financial base of local authorities. The financial base is the emanation of a community's economic base. The more diversified a community's economic base, the greater the financial resources mobilized to meet its operating and investment needs. Some elected representatives have taken steps to increase their own resources. But to date, they have not been able to exploit all the opportunities that exist in their local area.

13.3 Specific local development funds

Financing the development of local authorities by setting up specific funds appears to be one of the most credible alternatives available to local councillors. As part of our documentation on the issue of financing local authorities, we read: "*Our small budgets push us to innovate*". This statement was made by Jaime Lerner, architect and urban planner, former mayor of the city of Curitiba, in the Brazilian state of Paneras. During this mayor's three four-year terms in office, Curitiba made considerable progress in all areas. Everything that had been achieved was based on major innovations that were hailed by the major organizations involved in international cooperation for local development. Fascinated, they accompanied the ambitious municipal team in its development programs and projects. Curitiba was seen as a model of innovation and governance.

All local councillors in Côte d'Ivoire deplore the inadequacy of the financial resources granted by the State to its local authorities. This is the reality, and it's nothing new. This situation, which is the result of the aforementioned "unicité de caisse", prevents them from properly carrying out identified development projects. In some cases, they can't even get projects off the ground. But, at the same time, this situation should encourage local authorities to innovate by proposing mechanisms likely to generate additional resources to devote to investment.

The president of the Sud Comoé Regional Council, Docteur Aka Aouele, had initiated, in 2014, a development fund with popular participation. This fund, called "*Sud Comoé Invest*", is to be fed by shares with a minimum value of 1,000 CFA francs. Local elected representatives in Côte d'Ivoire should follow in the footsteps of the President of the Conseil Régional du Sud Comoé and others who have already taken initiatives to mobilize financial resources internally. Good examples can be emulated. However, setting up a specific fund in a local authority must be preceded by in-depth, strategic reflection on key aspects in order to avoid obstacles. At present, there are few examples of this in Côte d'Ivoire. This shows the scale of the challenge.

13.3 Financing through partnership development

The issue of financing local authority development can be addressed by international organizations. Indeed, given the current limits of public funding for local development and the lack of experience in promoting bold initiatives to finance local authorities, partnerships appear to be a promising avenue. However, it has not yet been sufficiently explored by local authorities in Côte d'Ivoire. This situation can be explained by a number of factors:

- Ivorian local authorities are unfamiliar with the procedures for developing this type of partnership. Local authorities therefore lack the know-how to develop win-win partnerships;
- The long time it takes to set up partnerships, to the point where most local councillors don't dare get involved. They believe, quite wrongly, that their work will only benefit their successors. Is this not a sectarian, developmental mindset, especially as administration is a continuum, and one works not for oneself, but for the community?
- Insufficient knowledge of the financial potential of local authorities makes it impossible to determine their real capacity to repay the loans to be applied for and, consequently, to justify their solvency;
- The traditional prudence of Côte d'Ivoire's supervisory authorities, who do not systematically provide the political support needed by local

authorities wishing to develop partnerships with external institutions. Government authorities, backed by the track record of other local teams, do not want local elected representatives to embark on a cycle of uncontrolled indebtedness.

Stanislas Zézé, CEO of the Bloomfield rating agency says, at the August 8, 2018 signing of the agreement with Aka Aouelé, President of the Conseil Général du Sud-Comoé: "*The aim is to encourage, more and more, local authorities to raise funds on the capital markets to enable them to make lucrative investments in their local authorities. And above all, to have a development that is not essentially based on state resources which, very often, are not sufficient*" . [9]

Money is everywhere and in large quantities on the international market. This market offers great prospects for financing local development. What is most lacking, according to international finance specialists, are projects, good projects, projects that are well set up or well put together - in short, bankable projects. Ivorian local authorities as a whole have not yet acquired the capacity to set up projects of this kind. They are free to call on the services of consultants to help them set up projects and attract funding. The mobilization of external financial resources is carried out by a very small number of Ivorian local authorities.

In view of the above, it's safe to say that there's still work to be done to popularize this type of partnership among local authorities, to enable them to secure significant funding. To do so, they need to be more aggressive, more ambitious, in order to capture greater financial resources.

This is an area worth exploring, because it could provide local authorities with substantial funding to carry out even more ambitious, results-oriented projects.

---

[9] Emeline Péhé Amangoua (2018), Le Sud-Comoé et l'agence Bloomfield sign une convention, in Fraternité Matin of August 11 and 12, 2018, page 9.

Chapter 14: Citizen participation in local development

It cannot be said often enough that citizen participation is the key to the success of any local development project. In Côte d'Ivoire, citizen participation in public action remains notoriously low. This participation must be more active, more offensive and more sustainable. This requires the implementation of reforms, particularly institutional reforms, to boost citizen participation in Côte d'Ivoire.

14.1 Empowering local populations

People are a powerful factor in development when they take part in development projects or actions. The empowerment of populations depends first and foremost on individual and then collective awareness. At the moment, people tend to confine themselves to making social and economic demands, without really taking responsibility.

Citizen participation in local development is the weakest link in the local development chain in Côte d'Ivoire. The observation is unanimously made. Generally speaking, this observation is not followed by concrete proposals. We need to go beyond this observation and propose a structured approach to make citizen participation effective and sustainable. This will help to revitalize political, social and economic life in our communities.

Citizen participation in community development is an absolute necessity. The population must play a decisive and active role in the local development process. But today, citizen involvement in development initiatives is sorely lacking. The citizen leaves all initiative to central or local public authorities.

In the context of local development, the citizen has the opportunity to learn at school, in his family, in his community, in the company or in any other organization where he finds himself. However, in each of these human organizations, they don't always have the opportunity to find out what "local development" is. Local development is not taught.

The village is the last link in the administrative structure of Côte d'Ivoire. The cleaning of a neighborhood or village is, par excellence, a matter of collective interest. This activity requires the mobilization of local manpower. It doesn't necessarily require money, except for the purchase of a few work tools (hoe, machete, rake, gloves, wheelbarrow). It requires leadership from the person or resident with the right profile.

14.2. Citizenship and fiscal responsibility

It cannot be stressed enough that paying taxes is a civic duty. Every citizen, whoever he or she may be, has a duty to pay taxes. The tax paid by the citizen entitles him or her to social and economic services, in quantity and quality. In Côte d'Ivoire, when these services are not provided, or are insufficiently or only partially provided, citizens have no opportunity to express their dissatisfaction. In fact, there is no channel through which they can make their voices heard.

The E-commune[10] or local development tool was proposed by the Comité National de Télédétection et d'Informations Géographiques (CNTIG) in 2013. It is being implemented in the pilot communes of Fresco, Grand-Bassam and Marcory. E-commune," says Dr Edouard Fonh-Gbei, Secretary General of the Comité National de Télédétection et d'Information Géographique (CNTIG), "*is a response to the problem of mobilizing resources in communes. It will help communes that adopt it to better plan local development. It will help to secure the budget by helping to reduce fraud and corruption. E-commune enables communes to identify their fiscal potential and better organize tax collection*.

14.3. Settling citizens in their territories

Citizens who leave their homeland are looking for places where they can lead a decent life, politically, socially, culturally, environmentally, economically and in terms of security. They leave their region or country of birth to go, or hope to go, to another region or country offering acceptable or even better prospects.

The massive departure of a local authority's workforce to other regions, or even other countries, is a loss for the sending region and obviously a gain for the receiving region. For the labor-sending region, it is difficult to build and implement a sound local development policy. The endogenous, self-sustaining development of a community must be achieved with all its long-term residents. It's imperative to put in place strategies aimed at securing populations in their terroir.

Young people are the human potential on which the quest for local development must be based. But this productive force is not sufficiently engaged in the fight for local development, either because of a lack of specific training, or because of incomplete training that is, above all, ill-suited to the needs of the community. This explains and justifies, to a certain extent, youth unemployment and, at the same time, raises the crucial question of the employability of graduates.

Local authorities should not be afraid of young people. They can spearhead political and economic action at community level. They must include them at the heart of their policies. Failure to do so could prove dangerous or even suicidal for the future of their community. Local councillors need to get to grips with the issue of young people.

14.4 Multiple players and communication

Today, the world has become a global village. At the level of each country, there are a number of players involved in local development. Indeed, local development calls on a diversity of stakeholders. The concept of "stakeholder" is not new, even if new actors are emerging. It involves all the social, cultural and economic players who can be mobilized to ensure the success of a development project or action in a given locality. The diversity or

---

[10] "Développement de proximité - Le projet e-commune a commencé", Anoh Kouao, Fraternité Matin of September 26, 2013.

multiplicity of stakeholders means that access to information requires a good communication system to reach them, individually and collectively.

Mobilizing the various stakeholders in the definition, preparation and implementation of local development initiatives requires not just a system, but also effective communication strategies. Each stakeholder needs a specific communication strategy.

Today, however, there is a serious lack of communication and, above all, dialogue between local elected officials and the various segments of the population. This lack of dialogue prevents the cohesion needed for local development.

14.5. Training citizens in democracy.

For citizens to become true democrats, they need to be well trained. Untrained or poorly trained citizens are an obstacle to the establishment of democracy at local and national level. Well-trained citizens, on the other hand, are willing to participate in democratic debate and thus play a role in the community's development process. This training can be provided in a variety of ways: self-training, academic training (for those following school or university courses), training by organizations or associations promoting democracy, and training by political parties, particularly for their activists. Today, political parties are not capable of training citizens to discern between the particular, the individual and the collective. What's more, democratic life, as advocated and practiced in Côte d'Ivoire, is rife with lies, cheating and a great deal of moral and intellectual dishonesty among the population.

14.6. Citizens' right to development

The quest for development is a right. The right to development is a universal right. Today, there is no structured demand from citizens and communities for development addressed to local elected officials. Certainly, citizens talk about development issues among themselves, generally in living rooms, bars, maquis or restaurants. Each one does so according to his or her understanding, emotion and, above all, political, social or economic position. But there is no structured community demand for local development. Nor is there any structured debate on local development at community or local level. This lack of debate at local level means that local people are unable to understand local development.

Today, a large proportion of local populations remain in the dark. Ignorance is the worst enemy of development, which is then hard to build. Here, too, efforts must be made to raise the level of local knowledge. This will enable local authorities to make real progress.

Chapter 15: Leadership for local development

In its march towards progress, any organization or community can count on men or women who stand out from the crowd. They have a certain edge over the other members of the community. Without necessarily being exceptional men or women, they nevertheless emerge from the pack, naturally or according to circumstances. These are the men and women to whom we attribute the Anglicism "leader", which literally means "driver". A leader is someone who has the responsibility and ability to lead a group or community to achieve its objectives.

15.1. Leadership and local development

Leadership is an essential part of running a community. To be truly beneficial to the community, leadership must be exercised with great humility. A leader's humility makes it easier to rally around him or her, and to manage relationships with other members of the community. Lack of humility sets the stage for the emergence of personal conflicts within a group. The leader must act with finesse and flexibility. Pride creates a gulf between the leader and the other members of the group, the components of the community.

Leadership requires respect for colleagues, collaborators and partners. It is on the basis of this consideration that community members can submit to each other and work together to solve the challenges facing the community.

It is also worth noting that the decentralization system in Côte d'Ivoire does not oblige local elected representatives to achieve results. As a result, some of them do not provide themselves with the resources and strategies they need to put their communities on the right track.

15.2. Leader or conductor

As a conductor, the local leader must be above the fray. He or she must rally the members of his or her team to exploit every development opportunity that presents itself. Indeed, economic development opportunities exist in all local communities, contrary to popular belief. But these opportunities are, for the most part, ignored or neglected. This means that preparatory work needs to be done to make local development players aware that opportunities are at hand. All they need to do is show imagination and innovation to find lasting solutions to problems initially considered insoluble.

15.3. Local leader commitment

The commitment of the elected official or local leader is of the utmost importance in carrying out his or her mission. The committed elected official fights, body and soul, almost daily for his or her community. However, most elected officials arrive at the head of their communities without necessarily being motivated by this desire for commitment.

It's also worth taking a closer look at the conditions under which elected officials arrive at the head of certain communities. Voting in local elections is not always carried out regularly, i.e. transparently. In some localities, voting has been marked by irregularities, intimidation and fraud, sometimes involving ballot box stuffing and multiple voting.

The process of drawing up electoral lists and the voting that follows are generally described as "transparent" by the organizations in charge of elections. In reality, voting is not always transparent, because transparency doesn't suit those who, in a normal election context, would have little chance of winning anything.

Moreover, in many local authorities in Côte d'Ivoire, the elected representatives have only one leitmotiv: to represent the party in the national political landscape. For these local councillors, it's all about increasing their party's weight on the national political scene. This single-minded approach to the role of local councillor is not conducive to progress or development in the community. This is not the mission of local authorities. Their mission is to create the conditions for improving people's living conditions.

For another category of elected representatives, it's their personal promotion on the national political scene that preoccupies them. They are obsessed and thirsty for "recognition". Promoting the quality of community governance, with the aim of improving the well-being of the population, is not necessarily the goal of these elected officials. The national context is marked by the absence of a genuine culture of excellence on the part of local councillors to offer the populations of their communities the best living conditions and frameworks.

15.4. Leader's commitment to young people

Côte d'Ivoire's local authorities have a large, young population whose talents and genius have not yet been sufficiently exploited. Young people are not sufficiently prepared to play their part in running community affairs. Young people in communities are not taught to work with their hands, with their heads, so as to exploit the opportunities that exist locally. More specifically, we need to teach them to :

- Respect the old and the new;
- To know themselves, to know the potential that exists within them;
- Identify local opportunities for social and economic integration;
- Work hard and hope to reap the rewards when the time comes;
- Build teams and get them to work optimally;
- Working together (it's a spirit) to achieve great things;
- Be patient, don't rush into anything, and wait confidently for the right moment (patience is always the golden way);
- Better manage financial resources, generated individually or collectively, when business is successful;
- Exercise their right to make their case to elected representatives.

15.5. Commitment to promoting local culture

Culture is the very foundation of development. Any development policy, program or project that is not based on local culture is doomed to failure. Local development, or the quest for it, must be fundamentally based on the community's cultural values.

Cultural diversity, which characterizes all local authorities in Côte d'Ivoire, is an asset in the quest for development. It can be the foundation on which everything is built. Cultural diversity is an element of mutual enrichment.

15.6. Commitment to fighting unemployment

Unemployment is one of the major challenges facing local authorities. The fight against unemployment or underemployment can be pursued in two main directions.

The first direction is for local councillors to offer direct employment in their departments. However, we must not delude ourselves, as their room for manoeuvre is limited by the organic framework of local authority jobs.

The second direction calls on local elected representatives and their partners to put in place the conditions for the creation of income-generating activities or businesses by local populations. Opportunities for economic integration exist in local communities. But due to ignorance, lack of ambition and, above all, lack of support, these opportunities are not sufficiently exploited. The conditions that need to be created or put in place include administrative and legal support, direct financing and tax benefits to enable fledgling businesses to grow and reach the required maturity. It also involves technical and managerial support for young local entrepreneurs. Local elected representatives need to find the synergies needed to support the actions of the various organizations involved in running the community's affairs.

15.7. Commitment to combating urban insecurity

Insecurity is an obstacle to local development in both urban and rural areas. It is essentially the result of the poverty or precariousness in which a large number of citizens, mostly young people, live today. This poverty is exacerbated by the lack of economic prospects for young people in these communities. Faced with the impossibility of satisfying their basic needs, many young people turn to illicit activities, which justifies the rise in insecurity.

The foundations for sustainable local development cannot be laid unless the social and economic environment of communities is secure. It is therefore essential to create optimal conditions for the security of people, goods and investments within communities.

15.8. Economic development cooperation

15.8.1. Current form of cooperation

Economic cooperation will continue to be practiced between nations, in order to bring the less advanced to benefit from their experience in terms of advice-assistance, technical or financial assistance. Over the years, however, cooperation has come up against its limits, with budget restrictions and cuts in particular. In its current form, this cooperation raises a number of questions.

We also note that international development aid organizations have "their own logic and procedures"[11] . The fact that they do not master their own logic and procedures does not make the task of poor nations any easier. This situation is a handicap to better exploitation of existing financing opportunities. In the context of economic cooperation, difficult, even draconian conditions are imposed on poor countries. Rolf Traeger continues: "This poses a problem in terms of transparency, monitoring the effectiveness of aid, and consistency with the objectives defined by each government".

15.8.2. Towards cooperation with local authorities

Reforms are increasingly required to make assistance to local authorities more productive, and therefore more beneficial to the local population. For example, international organizations are working directly with local authorities to better target the needs of their populations.

Cooperation between local authorities and international organizations focuses on programs or projects. This cooperation is based on a contract. The framework thus established defines the conditions for managing the partnership. This raises the question of technical and strategic support for local authorities. For greater efficiency, this technical support should be provided under the direction of national experts.

Cooperation is multi-faceted and multi-sectoral. It has not always benefited populations, because there is a gap between what poor countries consider good for their populations and the programs or projects financed by international organizations. This raises the question of the design of activities between the two levels.

15.8.3. Cooperation constraints

When it comes to cooperation with international organizations, local authorities in Côte d'Ivoire are not always in the best position. They are handicapped by both internal and external constraints. Internal constraints to the development of cooperation with international organizations include a lack of boldness, imagination and innovation on the part of some local elected representatives. Added to this is the absence or inadequacy of qualified human resources to cover their needs. There are also insufficient or weak financial resources available to local authorities. All these factors prevent development projects from being carried out.

For international organizations, the difficulties stem from a lack of understanding of local authorities. Indeed, local authorities have their own specific social, cultural, economic and environmental realities. Cooperation with local authorities is like entering another world. To penetrate this environment, international organizations must seek to offer appropriate cooperation.

---

[11]Rolf Traeger, head of the Least Developed Countries (LDC) section in the 2019 report of the United Nations Conference on Trade and Development (UNCTAD), published on November 19.

15.9. Technology transfer and local development

15.9.1. Technology transfer and local authorities

For development purposes, the question of technology transfer still arises today for nations, local authorities and companies. It remains a topical issue, despite the efforts and progress made by human organizations. There are several aspects to technology transfer. First and foremost, it should be pointed out that technology transfer does not simply mean the transfer of technology from the countries of the northern hemisphere to those of the southern hemisphere. For the countries of the South, technology transfer does not just mean the transfer of knowledge and know-how from national technical elites to grassroots communities. Rural or indigenous communities have been able to acquire technologies to solve the existential problems they face.

15.9.2 Clean or green technology

Generally speaking, when we speak of technological development, we see the effects of human action on nature. This tendency is mitigated by the use of green or clean technologies, which help to maintain, protect and promote nature. Green or clean technology is used to reduce the negative impact of human activity on the environment.

In the field of the physical environment, green technologies include waste recycling, wastewater treatment and renewable energies, particularly in the face of dwindling fossil fuel sources. Numerous other green or clean technologies are used in the building industry, transport, agro-industry, etc.

The quest for local development must be based on constant innovation. Through social, scientific or technological innovation, we seek and find solutions that are original, inexpensive, controllable, sustainable and, above all, adapted to the local context. Today's scientific and digital developments offer unprecedented opportunities for innovation in the field of local development management, insofar as they enable us to improve the quality of services and, by extension, the quality of life of local populations.

15.9.3. Local technology absorption capacity

One of the key elements in the question of technology transfer is the local or community capaci8ty to absorb the transferred technologies. More precisely, it is the technical capacity of local or community players to consume or use the technology received to solve the problem posed. Once the technologies are in hand, it's a question of seeing the technical capacity of community players to appropriate and use them to effectively solve the problems posed.

It is therefore necessary to assess this capacity beforehand, to determine whether it is sufficient or insufficient. In the event of insufficient capacity, local or community players need to be trained and empowered. Technology leaders need to be trained within communities. They will then teach other members of the community to master the technology.

15.9.4. Examples of local innovations

- ***Sandbag technique***

This Japanese technique[12] , tested in a pilot application in Anyama and Grand-Lahou, was presented to mayors from Côte d'Ivoire[13] .

- ***Making paving stones from plastic bags***

The young workers of the Plate-Forme des Services (PFS), Agboville section, have produced paid paving stones from plastic bags. These pavers are of good physical quality. However, the strength of this material needs to be tested, as it could be used as a substitute for bitumen in both urban and rural roadways.

The large-scale manufacture and use of this paving stone could solve the environmental problem of Côte d'Ivoire's towns and villages, which are overrun and ugly with plastic waste. Côte d'Ivoire is struggling to collect plastic waste from urban and rural public areas. With this outlet, plastic waste would be a major source of jobs and income for all those interested in it.

In conclusion, it's worth pointing out the importance and necessity of leadership development in the mental, social, cultural, economic, technological and environmental transformation of local communities. Good leadership makes it possible to mobilize human, financial and technical resources at local and community level, and to devise strategies for conducting activities in a coherent and effective manner.

---

[12] Professor Kimura Makoto (professor of geotechnics, geophysics and civil tunnel engineering) from Kyoto University presented this material to the mayors of Côte d'Ivoire.

[13] Franck Zabgayou (2015), Développement local - La technique des sacs de sable pour des routes à moindre coût, in Fraternité Matin August 21, 2015.

Chapter 16: Acting for sustainable local development

Côte d'Ivoire is looking for policies, programs, projects and strategies to ensure its sustainable progress. To achieve this ambition, the country must exploit all existing and future opportunities to lay the foundations for local and national development. This chapter outlines the specific measures required to achieve sustainable development.

16.1 Setting local development in motion

From our point of view, the quest for local development in Côte d'Ivoire has stalled or even stalled. The situation is clear, with the deplorable shortcomings in the provision of essential services and opportunities to the population, particularly in the sixteen (16) areas of transferred competence or responsibility. The engine of local development in Côte d'Ivoire has broken down. This is not a one-off breakdown. Rather, it is structural. It therefore appears necessary to rebuild the cards structurally in order to revitalize the process.

The mentality of the vast majority of Ivorians is that everyone wants a piece of the cake, or the nation's recognition, through an appointment to a public or private position, or through financial compensation. It is this generalized attitude that exacerbates political struggles or struggles for access to state power.

Citizen participation in local development or in the quest for local development should not be subject solely to questions such as: "*What's in it for me?* Rather, it should be based on questions such as: "*What can I contribute to the community*? It's by contributing, by giving something or by giving yourself to the community that you gain, yourself. But most people want to receive without having contributed to the community, and act to enable the local community to move forward. We need to reverse our way of thinking.

16.2. Prospects for real development

The development prospects of Ivorian local authorities were analyzed and discussed at the seminar to launch the process of diagnosing the capacities of local authorities (197 communes, 35 regions, 2 districts) organized on October 19, 2017 by the National Secretariat for Capacity Building. On this occasion, the following objectives were analyzed by the participants:

- Fixing populations by creating the conditions for their development. Fixing populations is not an empty word. It is, above all, a state of mind and an attitude to be adopted to enable people to stay in their region and undertake income-generating activities to improve their standard of living and well-being;
- Transform municipal waste to improve urban sanitation, increase revenues and create jobs, particularly in urban and peri-urban agriculture;

- Identify, prepare, implement and manage structuring projects with local populations. No project should be undertaken without involving the local population. Their contribution is crucial to the success of the project;
- Transforming local economies, in the knowledge that transforming a local economy is a process. It requires technical, technological and even strategic choices to be made;
- Develop local authorities' ability to seek financing;
- Training local players ;
- Technical support for local authorities;
- Strengthen local authorities' information capabilities.

16.3. Social justice

Social justice is one of the goals of local government action. They must create the conditions for social justice. Social justice is not an abstract concept. It must be a concrete reality. It means guaranteeing equitable access to social and economic resources and services for the community's populations, without any form of discrimination based on race, ethnicity, religion, region, social category, gender or any other factor. Social justice is the opposite of inequality in all its forms. It must be the concern of local elected representatives and their partners.

16.4. Need to change course

Côte d'Ivoire's local authorities are not moving forward, either stagnant or lethargic. In opting for decentralization, the public authorities decided to make Côte d'Ivoire's development dependent on the smooth running of the local authorities that make it up. However, these local authorities are not functioning properly, in such a way as to create the conditions for their development. They can no longer move forward unless a major change of direction is introduced. They need to be revitalized, to be shaken up a bit.

Local authorities in Côte d'Ivoire lack human and financial resources, as well as essential services such as water, electricity, health and education, in terms of both quantity and quality. This situation is generally justified by a lack of money. At the same time, we see an infinite number of economic activities that can be carried out on the basis of these services. The absence or inadequacy of basic services limits or even blocks creativity and imagination.

In the context of the quest for local development, we need to change course. We need to turn the page on the options that have been favored up until now, and which have failed to create the conditions for sustainable progress in our country. For a real change of course, we need to :

- Put an end to arbitrary, political territorial divisions that are incapable of laying the foundations for real, self-sustaining local development;
- Change the status of local councillors, transforming them from part-time employees into full-time "local" company directors, with a relatively competitive salary package, and the possibility for some to receive an annual bonus based on their team's performance. This will create a certain degree of emulation among elected representatives;

- Introduce a residency clause as a sine qua non condition for running for local elected office;
- Put an end to elected officials holding concurrent public or parapublic office. The function of local elected representative cannot be combined with another public or semi-public function. It is a demanding role. It must therefore be carried out on a full-time basis.

16.5. Obligation of Boards of Directors to deliver results

Increasingly, local councillors are being asked to manage local authorities like industrial or commercial enterprises. We're talking about the governance of companies and local authorities. Large public and private companies have boards of directors. Local authorities must have boards of directors. It is the Boards of Directors who, by controlling their activities, will lead local authorities to put in place the conditions to meet the obligation of results. The question that arises, then, is who should be on the Board of Directors. It should include representatives of the key institutional players: the Ministry of the Interior, the Ministry of the Economy, the Ministry of Construction and Urban Planning, the Ministry of the City, the local private sector (formal and informal), representatives of the supervisory ministry, a representative of civil society, etc.

16.6. Obligation to reside in the community

The chronic absenteeism of local councillors in Côte d'Ivoire is largely due to the fact that almost all of them live outside their local authority. Many of them live in Abidjan, the economic capital. Most of them only visit their local authority occasionally. This absenteeism poses problems for the elected representatives themselves, for the local authorities and for the local population.

Absent from their communities for days, weeks or even months at a time, elected representatives are unable to carry out their duties effectively. The absence of elected representatives prevents them from experiencing the problems facing the community. Over time, they become disconnected from these problems. Similarly, in the absence of elected representatives, local authorities in Côte d'Ivoire have difficulty signing and managing cooperation agreements with partners both inside and outside the country. Increasingly, there is talk of governance of local authorities, in the image of companies. Absenteeism removes a certain substance from this governance. It conveys to the public an image of elected representatives' lack of interest in their communities.

Furthermore, the prolonged absence of elected representatives from their communities is a source of frustration and even discontent for the local population, voters and non-voters alike. People don't see the people they elected. In the end, most people regret the choice they made in electing someone to head their local authority.

In view of the extent of absenteeism among local elected representatives, it seems necessary to make it compulsory for all candidates in local elections to reside in the locality. Côte d'Ivoire has trained managers in many areas of human activity. It is easy to find local people who are willing and able to lead local authorities towards progress.

16.7. The need to assess community performance

Assessing the performance of local authorities is a necessity. It aims to determine whether the objectives assigned to local elected representatives are being met. Performance measurement shows the extent to which a result has been achieved. A local authority's performance can be mediocre, weak, average, good or excellent. The idea behind assessing local authority performance is to encourage all local authorities to continually improve their performance, in order to move towards excellence.

In a context where the performance of local authorities is assessed, each local team gives its best in the various services provided to the population. People know how to appreciate the quality of the services they receive.

The purpose of the community performance assessment is to let elected representatives know that they are in competition. At the end of the assessment, the best-performing local authorities receive laurels or awards. The worst performers are content with encouragement from the public authorities to do more, to do better.

Lastly, local development partners use the results of local authority performance to better guide technical assistance programs or projects, so as to raise the level of their development.

16.8. Inclusive and equitable access to land for young people

Young people, a community's economic production force, have land needs to satisfy. They want to develop agricultural and pastoral activities, as well as leisure areas. They are also demanding the possibility of organizing agricultural processing activities, through small industrial units inside or on the outskirts of towns. In their search for solutions to their problems, young people generally come up against the indifference of local landowners.

In all cases, the issue of young people's access to land resources in local communities needs to be given careful consideration. It must be a priority for local teams.

16.9. Mobility within local authorities

The mobility of people within or between local communities is of obvious interest. It is a factor of social and economic dynamism. In terms of mobility, local elected representatives must commit to solving the following problems:

- Inadequate organization of the transport system for goods and people ;
- Inadequate means of transport in town and country;
- Rapid deterioration of road infrastructure within and between communities due to lack of maintenance;
- Attacks by citizens on road infrastructure, particularly in towns, hastening its deterioration.

16.10. Union des cadres de la collectivité

Local communities are plagued by a lack of unity and solidarity between populations. Poverty and precariousness are the main causes of difficulties within local communities. The battle for local community development is an uphill one. No matter how strong and capable an elected official or local leader may be, he or she, or his or her team, cannot lead the community along the road to progress. The battle against the underdevelopment of local government in Côte d'Ivoire can only be won by uniting men and women. This is eminently important.

Today, however, unity among the population is a rare commodity. The division between the community's political and economic leaders, on the one hand, and their supporters, on the other, has become profound since the last elections. As a result of this division, community members' efforts in favor of development are scattered and, consequently, ineffective. The community's leader or elected official must make the union or rallying of executives and populations one of his or her main hobbyhorses.

The achievement of union comes in a context in which everyone distrusts everyone else, in which no one trusts anyone. Moreover, there is more arrogance and pride than humility. Internal quarrels, interpersonal conflicts and low blows are worrying aspects.

Finally, in this difficult context, the role of the elected official or leader is above all to convey good ideas, ideas that bring people together, from generation to generation, in order to consolidate the community's achievements.

16.11. Bolder, more sustainable

The lack of audacity of Ivorians, particularly those who have the financial resources to carry out large-scale or ambitious projects, and who fail to do so. They have passed on this lack of audacity to the younger generations, who are also content with little. The lack of boldness shown by the sons and daughters of the locality has prevented it from advancing along the road to progress. This trend must be reversed if we are to improve the results achieved in the running of local government affairs.

In the same vein, and particularly with regard to choices in the consumption of goods and services, there is a frantic pursuit of luxury among economically successful leaders: cars, houses, furniture, travel, fashion and so on. They don't make enough productive investments that might lead them to leave their relatively good social and financial situation behind, and move towards a higher, more sustainable level of performance.

Today, objective observation of the results achieved through decentralization in Côte d'Ivoire can hardly encourage complacency. Indeed, the picture briefly painted in this chapter is not a glowing one, given the number of weak points to be improved in our decentralization system in Côte d'Ivoire. Despite this, there is a feeling of optimism within us that we can do better, that we can move forward with commitment and a willingness to overcome the obstacles that have and will continue to stand in the way of progress. Progress is possible, if each and every one of us, as far as he or she is concerned, is more daring every day in the pursuit of sustainable results, both for today's generation and for tomorrow's.

Conclusion

From 1960 to 1980, Côte d'Ivoire adopted a centralized approach to the management of state power. The State and its branches were the sole masters of the game. From that date onwards, the country decided to change course, with the introduction of municipalization into the power management system, with the direct participation of the population in the implementation of development programs, projects and actions. The choice of locally-based development for the nation had raised great hopes at all levels. Forty (40) years later, the expectations of public authorities, local elected representatives, local populations and development partners have been only partially met, if at all, in a large number of local authority areas. The results are average, if not disappointing. Faced with this situation, the essential question is what explains the lack of results from our locally-based development.

The despair of the population, the cynicism of central and local leaders, the lack of ambition of many citizens, widespread corruption, misinformation on all fronts, particularly via the notorious social networks, and the poor knowledge of the regions by local leaders, are the main reasons why Côte d'Ivoire is lagging behind. Added to this is the attitude of local leaders towards their populations, to whom they promise things they don't deliver. It's absolutely frustrating.

More objectively, it can be said that Côte d'Ivoire's semi-successful decentralization situation is the result of eight (8) factors: 1) the plethoric number of powers transferred to local authorities in Côte d'Ivoire; 2) the inadequacy of human and financial resources made available to local authorities; 3) the lack of openness of candidates in local elections, who promise the people great things but fail to deliver once elected; 4) the ignorance of the people, who are unable to discern between the good and bad programs proposed by candidates; 5) low popular participation; 6) poor knowledge of the realities of local government by candidates or elected representatives; 7) poor performance by local authorities in carrying out their mandates, due to the absence of any obligation to deliver results; and 8) absenteeism of local elected representatives, almost all of whom live outside their communities. The list is not exhaustive.

The march of a nation is like a chain. For the chain to move normally, quickly or optimally, its links must be able to move one after the other. The question today is how to improve the overall governance of local authorities in Côte d'Ivoire? Central public authorities, local elected representatives, local populations and local development partners all have their own opinions on the matter. It's not easy to find a consensus.

Among all the positions that can be defended by players from different groups, there is that of Ivorian experts in local development. The position defended in this book is, in fact, that of an Ivorian "expert". This position is that of a citizen who has not really been at the heart of local authority management. Is this an opportunity or a handicap? I consider it an opportunity, because I feel comfortable. I have no compulsion to defend the position that seems more neutral to me. I would point out that this position can, in many respects, be contested by other players, citizens or local elected representatives. By challenging this position, with supporting arguments, they will surely enable me to broaden and deepen my field of knowledge.

Bibliographic list Editions Foghel YAO NGoran

DAGBO NAGNON, Georges (2013). *Bilan diagnostic de la région du Haut Sassandra, 2010-2011, Daloa*, Direction régionale du Plan et du Développement, 2013, 116 p.

GARFIELD, Charles (1986). *Haute performance - La clef du succès en affaires,* Expansion /Hachette, 1986, 321 p.

JAGLIN, **Sylvy and DUBRESSON, Alain** (1993). *Pouvoirs et cités d'Afrique noire: décentralisation en question.* Editions Karthala, 1993, 308 pages.

JOYAL, André (2002). *Le développement local-Comment stimuler l'économie des régions en difficulté.*
IQRC Studies, 2002, 156 pages.

MOURAMANE, Fofana. *Rêver le progrès,* CEDA, 1997, 251 p.

PNDL(2011). Classification of the roles and responsibilities of decentralization actors. Final report, Plan National de Développement Local, Senegal, 2011.

BNETD (1998). *Guide des équipements communaux.* Bureau Nationale d'Etude Techniques et de Développement (BNETD), 1998, 376 p.

AMADOU, Diop (2008). *Développement local, gouvernance territoriale : enjeux et perspectives,* Editions Karthala, 2008, 230 pages.

BELLINA, Séverine (2008). La gouvernance démocratique: un nouveau paradigme pour le développement? Paris, Karthala, 2008, 608 p.

BEVIR, Mark (1996), "*Local governance*", United Nations Forum held in Gothenburg (Sweden) from September 23 to 27, 1996.

DECOSTER, Dominique Paule (2002). *Gouvernance locale, développement local et participation citoyenne,* Charleroi, November 2002, 96 pages.

DUBRESSON, Alain. and FAURE, André Yves. *Décentralisation et développement local : un lien à repenser.* Revue Tiers-Monde 1, N° 181, p. 7-20, PUF.

JEROME, Marie and IDELMAN, Eric (2010). Decentralization in West Africa: a revolution in local governance? EchoGéo, 13/2010 , July-August 2010.

ADAMOLEKU, Lapido (1989). *Decentralization policies in Sub-Saharan Africa: an overview.* Handout, 1989, World Bank, 67 p.

ATTAHI, K and YAO, Bazin (1999). *Urban development and good governance. Etude du cas d'Abidjan.* Unpublished paper, commissioned by Empirica Group (South Africa), 1999 ;

WORLD BANK (1995). *Pour de meilleurs services urbains- Trouver les bonnes incitations.* World Bank, 1995, 96 p.

ILO (2001). *The impact of decentralization and privatization on municipal services.* ILO, Geneva, 2001, 114 pages.

BOUINOT, Jean (2002). La Ville Compétitive - Les clés de la nouvelle gestion urbaine, Paris, Economica, 2002.

DUBRESSON, Alain and FAURE, Yves André. Décentralisation et développement local : un lien à repenser. Revue Tiers-Monde 1, N° 11, p. 7-20, PUF.

COMAQ (2000) Performance Indicators for Quebec Municipal Organizations. Corporation des Officiers Municipaux Agréés du Québec *(COMAQ),* research report on indicator development, 2000, 86 p.

GREFFE, Xavier (2002). *Le développement local.* Paris, éditions de l'Aube, Datar, 2002, 200 p.

IRFED (1996). *Trainer les élus et responsables locaux au développement local dans le contexte des décentralisations africaines.* Manuel à l'usage des formateurs et des formateurs de formateurs, IRFED, Paris, 1996, 346 p.

YAO Bazin (2019), Débat sur le développement local, Editions du CERAP, Abidjan, 263 pages.

YAO, Bazin (1998). Commercializing the city: the ignored dimension. Afriqurba. 1998, N° 004.

Dagobert Banzio (SG of Ardci) (2017), "L'Etat nous a transféré 16 compétences sans que les moyens ne suivent", in Fraternité Matin, February 28, 2017,

BOMBET, Emile Constant (1997). Exposé introductif présentant le programme gouvernemental de décentralisation et d'aménagement du territoire, à l'occasion de la Table Ronde des Bailleurs de Fonds sur la " *Décentralisation et l'Aménagement du Territoire* ", tenue à Yamoussoukro, du 12 au 14 Mai 1997.

Julien Denormandie (2020), French minister, in charge of cities, on the occasion of the Abidjan meeting, February 27-28, statement reported in the daily Fraternité Matin of February 28, 2020, page 3.

Emeline Péhé Amangoua (2018), Le Sud-Comoé et l'agence Bloomfield sign une convention, in Fraternité Matin of August 11 and 12, 2018, page 9.

Anoh Kouao (2013), "Développement de proximité (2013) - Le projet e-commune a commencé", in Fraternité Matin, September 26, 2013.

Printed by Books on Demand GmbH, Norderstedt / Germany